Polymers

Properties and Applications

7

Walter Klöpffer

Introduction to Polymer Spectroscopy

With 80 Figures

Springer-Verlag
Berlin Heidelberg New York Tokyo 1984

Prof. Dr. Walter Klöpffer
Leiter der Zentralstelle Qualität und Abt. Umwelttechnik
Battelle-Institut e.V.
Am Römerhof 35
D-6000 Frankfurt/M. 90

This volume continues the series *Chemie, Physik und Technologie der Kunststoffe in Einzeldarstellungen,* which is now entitled *Polymers/Properties and Applications.*

ISBN 3-540-12850-6 Springer-Verlag Berlin Heidelberg New York Tokyo
ISBN 0-387-12850-6 Springer-Verlag New York Heidelberg Berlin Tokyo

Library of Congress Cataloging in Publication Data:
Klöpffer, Walter, 1938– Introduction to polymer spectroscopy. (Polymers: properties and applications; 7) Includes index. 1. Polymers and polymerization–Spectra. I. Title. II. Series: Polymers; 7.
QC463.P5K55 1984   547.7'028   83-16700
ISBN 0-387-12850-6 (U.S.)

Typesetting, printing, and bookbinding: Brühlsche Universitätsdruckerei, Giessen
2154/3020-543210

This book is dedicated to

ERWIN SCHAUENSTEIN
Professor of Biochemistry
at the University of Graz

on the occasion of his 65th birthday.

Preface

This book has grown out of several courses of lectures held at the University of Mainz in the years 1978 to 1981, at the Ecole Polytechnique Fédéral, Lausanne, and at the University of Fribourg, Switzerland. The last two courses were held in the framework of the "3e Cycle" lectures in June 1981.

According to this genesis, the emphasis of the book lies on a unified and concise approach to introducing polymer spectroscopy rather than on completeness which, by the way, could hardly be achieved in a single volume. In contrast to other books on this subject, equal weight is given to electronic spectroscopy, vibrational spectroscopy and spin resonance techniques. The electronic properties of polymers have been increasingly investigated in the last ten years; until recently, however, these studies and the spectroscopic methods applied have not generally been considered as part of polymer spectroscopy.

The increasing use of electronic spectroscopy by polymer researchers, on the other hand, shows that this type of spectroscopy provides efficient tools for gaining insight into the properties of polymers which cannot be obtained by any other means.

Although the main aim of the book is a didactic one, it should also be useful as a first survey to polymer researchers not specialised in spectroscopy or even to specialists in one or the other spectroscopic technique wishing to orient themselves outside their own field of research. Much emphasis is therefore given to the understanding of the basic processes and mechanisms involved in each method, which are dealt with in an elementary fashion. Both differences and similarities between low and high molar mass compounds with regard to their spectra have been elaborated. The peculiar status of macromolecules, somewhere between molecule and one-dimensional crystal, is demonstrated in many instances. At the end of each chapter, the strength as well as the limitations of the method are delineated, showing that only the combination of several methods can yield an adequate picture of a given polymer.

This book would never have come to live without the initiative and constant interest shown by Dr. F. L. Boschke, Springer-Verlag. It was originally planned as a multi-authored volume; the help and encouragement I got in this early phase by H. J. Cantow (Freiburg), J. J. Verbist and J.-N. André (Namur), H. Kashiwabara (Nagoya) and R. Kosfeld (Düsseldorf) is gratefully acknowledged.

The English manuscript has been corrected and improved by Mrs. E. Zimmer, Battelle-Institut Frankfurt, in addition to her official duties. My sincere thanks are also due to U. Wegstein, M. Ilgenstein and G. Teichmann for typing the consecutive stages of the manuscript. Most figures have been drawn by Mr. O. Ruppel.

Most of my own research, which is quoted in Chapters 3 and 4, has been performed within the "Battelle Institute Programme in Physical Sciences".

Finally, I would like to extend my thanks to all colleagues who sent me reprints and manuscripts before publication, to A. Braun, Lausanne, for organising the guest lectures in Switzerland and, last but not least, to my students for their endurance and for their comments.

Frankfurt am Main, June 1983   Walter Klöpffer

Table of Contents

Part A. General Introduction 1

1   Introduction 1
1.1   Definition of Polymer Spectroscopy 1
1.2   Information Obtained by Polymer Spectroscopy 1
1.3   Spectral Range 3
1.4   Remarks on the Choice of Examples for Applications 4
1.5   Bibliographic Notes 4
        References 5

Part B. Electronic Spectroscopy 7

2   ESCA in Polymer Spectroscopy 7
2.1   Principle of the Method 7
2.2   The ESCA Spectrometer 8
2.3   Main Features of ESCA Spectra 9
2.4   Applications of Core-Electron Spectra 11
2.4.1   The Chemical Shift 11
2.4.2   Bulk and Surface Analysis 13
2.4.3   Examples of Bulk Analysis 14
2.4.4   Examples of Surface Analysis 14
2.5   ESCA Spectrum of Valence Electrons 15
2.5.1   Energy Bands 15
2.5.2   Energy Bands in Polymers 17
2.5.3   The Zero-Level Problem in ESCA Studies of Valence Bands 18
2.5.4   Examples of Valence Band Studies Using ESCA 18
2.6   Résumé of ESCA in Polymer Spectroscopy 21
        References 22

3   Absorption Spectroscopy in the Ultraviolet and Visible Regions . . . 23
3.1   The far Ultraviolet (fUV) 23
3.2   Absorption Spectroscopy in the Near Ultraviolet and Visible Regions . 25
3.2.1   Spectral Region and Characteristic Transitions 25
3.2.2   Basic Principles of nUV/VIS Absorption Spectroscopy 26
3.2.2.1 Electronic Vibronic Spectra 26
3.2.2.2 Energy Level Diagrams 27
3.2.2.3 Intensity of Absorption Bands 27
3.2.2.4 The Franck-Condon Principle 28
3.2.3   Application of Lambert-Beer's Law to Polymers 28

3.2.4   UV-Absorbing Polymers 30
3.2.5   Hypochromy in Polymers 32
3.2.6   Absorption by Molecules dissolved in Solid Polymers 33
3.3   Résumé of nUV/VIS-Absorption Spectroscopy 34
        References 34

4   Fluorescence- and Phosphorescence Spectroscopy of Polymers 36
4.1   Radiative and Radiationless Transitions 36
4.1.1   The Jablonski Diagram 36
4.1.2   Fluorescence 37
4.1.3   Internal Conversion and Inter-System Crossing 37
4.1.4   Phosphorescence 37
4.2   Experimental 37
4.3   Quantum efficiency, decay time and rate constants 40
4.4   Fluorescence in Polymers 41
4.4.1   Isolated and Crowded Fluorescent Groups 41
4.4.2   Fluorescence from Isolated Chromophores 42
4.4.3   Energy Transfer between Isolates Groups or Dissolved Molecules . . . 43
4.4.4   Singlet Excitons 44
4.4.5   Excimer Fluorescence as a Probe in Polymer Studies 46
4.5   Phosphorescence in Polymers 47
4.5.1   The Phosphorescent Triplet State 47
4.5.2   Isolated Phosphorescent Groups 48
4.5.3   Triplet Excitons in Polymers 49
4.6   Résumé of Fluorescence and Phosphorescence Spectroscopy 50
        References 51

Part C. Vibrational Spectroscopy 53

5   Vibrations 53
5.1   Introduction 53
5.2   The Harmonic Oscillator 53
5.3   Molecular Vibrations as Quantum Phenomena 54
5.4   General Remarks on the Interpretation of Vibrational Spectra 55
5.5   Symmetry and Fundamental Vibrations of One-dimensional Chain
        Molecules 56
5.6   Phonons 59
        References 61

6   Raman Spectroscopy 62
        Introduction 62
6.2   The Smekal-Raman Effect 62
6.3   Experimental 65
6.4   Examples of Laser-Raman Spectra of Synthetic Polymers 66
6.5   The Accordion Vibration 70
6.6   Resonance Raman Scattering 72
6.7   Résumé of Raman Spectroscopy of Polymers 74
        References 75

7   Infrared Spectroscopy of Polymers   75
7.1   Introduction   75
7.2   Absorption of Infrared Radiation   76
7.3   Experimental   78
7.3.1 IR-Absorption   78
7.3.2 IR Reflection and Emission   80
7.4   Interpretation of Polymer mIR Spectra   81
7.4.1 Empirical IR Spectroscopy   81
7.4.2 Examples of mIR Spectra of Linear Polymers   82
7.4.2.1 Poly(methylmethacrylate)   82
7.4.2.2 Poly(isobutene) and amorphous polypropylene   85
7.4.2.3. Poly(vinylalcohol)   86
7.4.2.4 Poly(acrylonitrile)   87
7.4.2.5 Poly(caprolactam)   88
7.4.3 Tacticity and Conformation in mIR Spectra   89
7.5   Applications of mIR Spectroscopy   94
7.5.1 General   94
7.5.2 Copolymers   94
7.5.3 Molar Mass   95
7.5.4 Branching   95
7.5.5 Carbon Double Bonds   96
7.5.6 Oxidation Processes   96
7.5.7 Corona and Plasma Treatment of Polymer Surfaces   98
7.5.8 Dissociation of Polymeric Acids and Other polymer-IR Studies   99
7.6   Résumé of mIR   99
7.7   Far-Infrared Spectroscopy of Polymers   100
7.7.1 Introduction   100
7.7.2 Chemical Applications   100
7.7.3 Phonons in fIR Spectra of Polymers   101
7.7.4 Inelastic Neutron Scattering (INS)   102
7.7.5 Résumé of fIR + INS   105
       References   105

Part D. Spin-Resonance Spectroscopy   109
8   Principles of Spin-Resonance Spectroscopy   109
8.1   Introduction   109
8.2   The Spin of Elementary Particles   109
8.3   Resonance Absorption   110
8.4   Spin-Resonance in Polymers   111
       References   113

9   Electron-Spin-Resonance (ESR) Spectroscopy of Polymers   113
9.1   General Characteristics of ESR Spectra   113
9.2   Experimental   120
9.3   Survey of Polymer-Specific Applications of ESR Spectroscopy   123
9.4   Polymerisation Studies by Means of ESR   123
9.5   Mechanically Formed Polymer Radicals   126

9.6   ESR of Radicals Formed by Radiation 129
9.7   Triplet States 130
9.8   Spin-Labels 133
9.9   Résumé 135
       References 136

10   Nuclear Magnetic Resonance (NMR) Spectroscopy of Polymers . . . 137
10.1   The Origine of NMR Spectra 137
10.2   Experimental 143
10.3   High-Resolution ^{1}H-NMR of Polymers 144
10.3.1 Applications 144
10.3.2 Tacticity Analysis 145
10.4   Broad-line NMR 157
10.5   Spin-Relaxation Times 159
10.6   Chemically Induced Dynamic Nuclear Spin Polarisation (CIDNP) . . 160
10.7   Résumé of ^{1}H-NMR Spectroscopy 160
10.8   Polymer NMR Spectroscopy of ^{13}C and Other Nuclei 161
10.8.1 Experimental 161
10.8.2 Solid Polymer ^{13}C-NMR Spectroscopy 162
10.8.3 High Resolution ^{13}C-NMR Spectroscopy of Polymer Solution . . 165
10.8.4 Other Nuclei 168
10.9   Résumé 169
       References 169

Part E. Conclusion and Appendices 171

Conclusion 171
Appendix 1: Table of Polymers 173
Appendix 2: List of Abbreviations 174
Appendix 3: List of Elementary Constants Used in This Book 179

Subject Index 181

Part **A. General Introduction**

1 Introduction

1.1 Definition of Polymer Spectroscopy

Polymer spectroscopy deals with the application of a wide range of spectroscopic methods to the study of polymers, i.e. substances of high molar mass which are formed by linking of suitable low molar mass compounds – the monomers. With regard to their origin, we can distinguish synthetic polymers and biopolymers, with regard to chemical composition between homopolymers and copolymers and between linear, branched and cross-linked polymers.

Because of the elevated molar mass some physical properties of polymers are drastically different from those of monomeric (or oligomeric) substances, so that not all methods available for the latter are applicable to the former; e.g., it is not possible to evaporate polymers without decomposition and therefore gas phase methods can only be applied in order to study small fragments formed in the pyrolysis of polymers. This will be outside the scope of this book whose focus is on methods giving information on integral macromolecules and their aggregates rather than to monomeric degradation products.

Spectroscopy is usually defined as the science of the interactions between radiation and matter and deals with both the experimental and the theoretical aspects of these interactions.

This definition is too broad to be applicable to our purpose without reservation. First of all, we shall restrict our interest primarily to electromagnetic radiation and its interaction with polymers. With regard to the nature of interaction we shall further restrict ourself to quantum resonance phenomena; accordingly, pure relaxation effects such as microwave absorption of polymers, will be outside the scope of this book. The same is true of "elastic" diffraction and scattering of radiation not connected with uptake or release of energy. Inelastic scattering, on the other hand, is part of spectroscopy within the scope of this definition, e.g. Raman spectroscopy.

Summing up, we define polymer spectroscopy as the science of quantum resonance interactions of electromagnetic radiation with polymers. This does not exclude the occasional discussion of effects due to particle radiation such as neutron radiation, if complementary, or otherwise inaccessible information can be obtained.

1.2 Information Obtained by Polymer Spectroscopy

The primary information of spectroscopy always consists in energy differences, supplemented by additional pieces of evidence such as line and band shapes, intensities, and in some experiments by the time dependence or polarisation of signals.

Table 1.1. Informations obtained by spectroscopy relating to structure and dynamics of polymeric systems

Structure	Dynamics
Chemical structure	Movements of the chain,
Tacticity	segments and side groups
Conformation	Phonons
Crystallinity	Excitons
Electronic structure	Complex formation and related phenomena

From this primary information, polymer spectroscopy provides two types of information, related to the structure and dynamics of polymers (Table 1.1). Information on chemical – or primary – structure is the main aim of analytical applications of polymer spectroscopy. This type of information is especially important if complex copolymers, resins, etc. have to be characterised or identified.

Tacticity or configuration of the macromolecules often determines the degree of crystallinity which can develop along the chain (1D crystallinity or – regular – chain conformation) or in three dimensions. Equally important as the arrangement of the atoms in space is the average distribution of electrons or electronic structure which can be explored by several spectroscopic methods.

The dynamic phenomena to be studied by spectroscopy encompass movements of atoms and larger parts of the macromolecules, e.g. side groups. If the movements are collective oscillations of the lattice – in crystalline regions – they are called phonons.

Excited states may also move, in general without a corresponding molecular diffusion, in the form of excitons. Finally, there is a wide range of more or less specific molecular interactions which can best be described as complex formation and can be studied by suitable spectroscopic methods.

Of course, not all spectroscopic techniques give information on all the items listed in Table 1.1 and some of these can be more conveniently studied by non-spectroscopic methods; as an example, (3D-)crystallinity can be determined much more directly using diffraction techniques. This inferiority of spectroscopy applies in general to the study of long-range phenomena, since spectroscopy, as a rule, is somewhat "short-sighted". This disadvantage is, on the other hand, an asset insofar as spectroscopy yields precise information at the molecular level. Influences due to surrounding groups are evidenced by – mostly weak – disturbances. Depending on the magnitude of the disturbances and the interactions causing them, we may distinguish two complementary aspects of polymer spectroscopy:
- the polymer aspect:
    here, the absorbing or emitting group is either the entire macromolecule (or crystal lattice) or part of it which measurably and specifically interacts with other parts
- and the monomer aspect:
    in this limiting case the absorbing or emitting group is part of the macromolecule without measurable interaction with neighbouring groups. The spectra are identical with those of suitable monomeric model compounds.

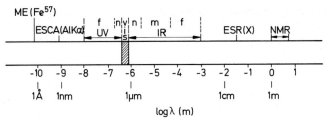

Fig. 1.1. Spectral range of spectroscopic methods used in polymer spectroscopy in a logarithmic wavelength scale. For converting photon energy (E), wave number (v') and wavelength (λ) of electromagnetic radiation it is advisable to remember the following approximate correlation:

Energy   wavenumber wavelength

1 eV/photon $\hat{=} 8065 \, \text{cm}^{-1}$ $\hat{=} 1240 \, \text{nm}$

($= 96.5 \, \text{kJ/mol}$)
(based on $E = hv = hc/\lambda = hcv'$)

Polymer physicists are in general more interested in the polymer aspect, whereas chemists and especially analytical chemists often use the addidivity of spectral bands inherent in the monomer aspect. Even here, it will often not be possible to avoid interactions completely, since polymers never form true dilute solutions with respect to the parts of the macromolecules, e.g. side groups in vinyl polymers.

As a practical consequence of this distinction it becomes necessary to study polymers and suitable monomers – which as a rule are not the chemical monomers – in comparison, the chemical nature of the model monomers depending on the spectroscopic method applied. Polymer spectroscopy therefore is not only the spectroscopy of polymers. It involves the investigation of monomers and oligomers, mostly in solution but occasionally also in the form of single crystals, polycrystalline or amorphous layers, thus offering many connections between polymer spectroscopy on the one hand and molecular and solid state physics on the other. In order to get a maximum of useful information, polymer chemistry and non-spectroscopic methods of polymer science have to be combined with polymer spectroscopy in order to produce pure and well characterised polymer samples.

1.3 Spectral Range

The spectral range covered by the different methods to be discussed is shown in Fig. 1.1. It extends at the wavelength scale used from about 0.1 nm (1 Å) to 10 m. Below this range, in the region of soft γ-rays, there are nuclear transitions showing the Mössbauer effect, a nuclear resonance absorption. Mössbauer spectroscopy which is based on this effect is unfortunately strongly restricted in its application to polymers due to the rare occurrence of heavy nuclei in polymers. Although γ-rays belong to the electromagnetic spectrum, Mössbauer spectroscopy will not be treated in this book due to the scarcity of examples.

The X-rays used for excitation in ESCA experiments have wavelengths near 1 nm. The experimentally difficult far UV (fUV) or vacuum-UV is found between 10 and 200 nm, followed by the important near UV (nUV) which forms a unit with the visible region (VIS) both with regard to the mechanism of electronic excitation and from the experimental point of view. It is the spectral sensitivity of the human eye which draws a borderline between nUV and VIS at about 400 nm. In this range, nUV- (and VIS-) absorption and luminescence (fluorescence and phosphorescence) of polymers have to be considered.

Next to the long-wavelength side we see the near infrared (nIR) from 0.8 to 2.5 μm, a much neglected part of the spectrum in polymer research, and further to long wavelengths, the most popular part (mIR) up to about 50 μm. This is followed by the far IR (fIR) comprising wavelengths up to about 1 mm. For theoretical reasons, Raman spectroscopy belongs to the IR range, although the excitation wavelengths used are mostly in the visible part of the spectrum, since the vibrational transitions observed in this inelastic scattering are similar to those observed in the IR spectra.

Further to higher wavelengths we find the relaxation spectra which do however not belong to the topic of this book, followed by ESR ($\lambda = 1$ to 3 cm) and NMR ($\lambda = 1$ to 5 m) spectroscopy.

The whole range can conveniently be divided into three parts:
- Electronic spectroscopy (ESCA and UV/VIS) in Part B
- Vibrational spectroscopy (IR and Raman spectroscopy) in Part C
- Spin resonance spectroscopy (ESR and NMR) in Part D

1.4 Remarks on the Choice of Examples for Applications

Polymer spectroscopy is an equally important tool for investigations of synthetic and biological polymers, both being highmolar mass substances and obeying the same chemical and physical laws. Although there is no theoretical reason for treating the two groups separately, there are limitations of space and editorial points of view that restrict the examples to the field of synthetic polymers. We nevertheless hope that the book can give some useful information for researchers in the field of biopolymers, the introductory and general sections being, of course, independent of the specific polymers to be studied.

1.5 Bibliographic Notes

The scientific literature on polymer spectroscopy has rapidly grown within the last few years so that complete coverage could only be tried in a several volume handbook and is outside the scope of this introduction. The books covering parts of the whole field are listed at the end of this chapter. The most comprehensive book on polymer spectroscopy is the one edited by D. O. Hummel. This excellent book can only be blamed for having been written at a time when the progress of electronic spectroscopy of polymers was not yet so clearly visible as it is today. We hope to fill this gap and to give a useful overview of the whole field of polymer spectroscopy within the framework of the definition given in Sect. 1.1. The quotations refer preferentially to books, reviews and important original papers. Since we shall also touch some very recent fields of research, in these cases also preliminary work will be quoted.

References

1. Hummel, D.O. (ed.): Polymer Spectroscopy, Weinheim: Verlag Chemie 1974
2. Iving, K.J. (ed.): Structural Studies of Macromolecules by Spectroscopic Methods, London: Wiley 1976
3. Jones, D.W. (ed.): Introduction to the Spectroscopy of Biological Polymers, London: Academic Press (1976)
4. Hoffmann, M., Krämer, H., Kuhn, R.: Polymeranalytik I and II, Thieme Taschenlehrbuch der organischen Chemie B4, Stuttgart: Thieme 1977; English translation: Polymer analytics, Polymer Monographs 3, Chur: MMI Press 1981
5. Brame, E.G. (ed.): Applications of Polymer Spectroscopy, New York: Academic Press 1978
6. Fava, R.A.: Polymers, Part A: Molecular Structure and Dynamics, Vol. 16 of Methods of Experimental Physics (L. Marton and C. Marton eds.), New York: Academic Press 1980
7. André, J.-M., Ladik, J., Delhalle, J. (eds.): Electronic Structure of Polymers and Molecular Crystals, New York: Plenum 1975
8. Natta, G., Zerbi, G. (eds.): Vibrational Spectra of High Polymers, J. Pol. Sci. Part C, No. 7, New York: Interscience Publ. 1964
9. Zbinden, R.: Infrared Spectroscopy of High Polymers, New York: Academic Press 1964
10. Siesler, H.W., Holland-Moritz, K.: Infrared and Raman Spectroscopy of Polymers, New York: Dekker 1980
11. Kinell, P.-O., Rånby, B., Runnström-Reio, V. (eds.): ESR Applications to Polymer Research, Stockholm: Almquist, New York: Wiley 1973
12. Rånby, B., Rabek, J.F.: ESR Spectroscopy in Polymer Research, Polymers – Properties and Applications, Vol. 1, Berlin, Heidelberg, New York: Springer 1977
13. Woodward, A.E., Bovey, F.A. (eds.): Polymer Characterization by ESR and NMR, ACS Symp. Ser. No. 142, Am. Chem. Soc., Washington 1980
14. Boyer, R.F., Keinath, S.E. (eds.): Molecular Motion in Polymers by ESR, New York: Harwood Academic Publ. 1980
15. Bovey, F.A.: High Resolution NMR of Macromolecules, New York: Academic Press 1972
16. Randall, J.C.: Polymer Sequence Determination (Carbon-13 NMR Method), New York: Academic Press 1977
17. Pasika, W.M. (ed.): Carbon-13 NMR in Polymer Science, ACS Symp. Ser. No. 103, Am. Chem. Soc., Washington 1974
18. Morawetz, H., Steinberg, I.Z. (eds.): Luminescence from Biological and Synthetic Macromolecules – Eighth Katzir (Katchalsky) Conference, Annals of the New York Academy of Sciences, Vol. 366 1981
19. Hummel, D.O. (ed.): Proceedings of the 5th European Symposium on Polymer Spectroscopy, Weinheim: Verlag Chemie 1979
20. Cantow, H.-J. et al. (eds.): Luminescence, Adv. Pol. Sci. Vol. 40, Berlin, Heidelberg, New York: Springer 1981
21. Phillips, D., Roberts, A.J. (eds.): Photophysics of Synthetic Polymers, Science Reviews, Northwood, England 1982
22. Painter, P.C., Coleman, M.M., Koenig, J.L.: The Theory of Vibrational Spectroscopy and its Application to Polymeric Materials, New York: Wiley 1982
23. Mort, J., Pfister, G. (eds.): Electronic Properties of Polymers, New York: Wiley 1982

B. Electronic Spectroscopy

2 ESCA in Polymer Spectroscopy

2.1 Principle of the Method

Electron Spectroscopy for Chemical Analysis (or Applications) – ESCA – is based on the following principle [1, 2]:
monochromatic short-wavelength radiation is used in order to detach electrons from the sample, the kinetic energy of the electrons being measured by means of an electron analyser. The difference in energy between the exciting photons of frequency v and the electrons detached (E_{kin}) is equal to the binding energy of the electron.

$$E_B = hv - E_{kin}. \tag{2.1}$$

The ESCA spectrum therefore informs us about the binding energies or ionisation energies of the electrons in the sample investigated. According to the Koopmans Theorem, the ionisation energies describe the sequence of atomic or molecular orbitals quantitatively. This is shown schematically in Fig. 2.1 for two orbitals 1 and 2 of different energy, separated by I_1 and I_2 from the zero level of energy, attributed arbitrarely to the detached electron without surplus kinetic energy and outside the attractive potential of the positive ion left during ionisation. The ionisation energies measured by ESCA are one-electron ionisation energies, not to be confused with the 2nd, 3rd ... ionisation energy of atoms, related to the formation of multiply charged ions.

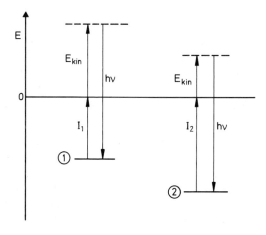

Fig. 2.1. Schematic energy level diagram for an ESCA experiment at constant exciting photon energy hv. Two orbitals 1 and 2 of different ionisation energy I indicated

B. Electronic Spectroscopy

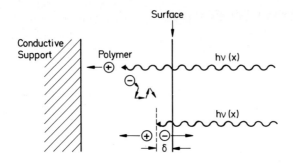

Fig. 2.2. Behavior of detached electrons in the bulk and at the surface of the polymer

Depending on the magnitude of $h\nu$, two experimental variants of electron spectroscopy can be distinguished:

1. excitation in the fUV ($h\nu \approx 20$ eV): this method is often called Photo Electron Spectroscopy (PES or UVPS) and so far has been applied only in few cases for the study of polymers [24].

2. X-ray excitation ($h\nu \approx 1,500$ eV): ESCA or XPS. Using this high energy excitation, even the firmly bound core electrons can be detached.

Applying ESCA to polymers, only solid state spectra can be obtained (in solutions the solvent molecules would mask the effect of the solute). In solids, the detached electrons are rapidly thermalised, at least in the bulk (Fig. 2.2) of the sample. Within a very thin surface layer, however, of the order of $\delta \approx 2$ nm, the electrons are allowed to leave the surface without loss in kinetic energy, so that Eq. (2.1) can be applied. Hence, ESCA is uniquely suited for studying the surface of solid polymers.

A serious experimental difficulty is posed by the extremely low electrical conductivity of most polymers ($\sigma < 10^{-15}\ \Omega^{-1}\ cm^{-1}$). Each emitted electron leaves behind a positive hole which cannot flow to the back electrode (shown in Fig. 2.2) fast enough in order to prevent static charging. The electrons emitted have to surpass this attractive potential and consequently lose part of their kinetic energy. This effect has to be compensated by suitable calibration (evaporation of gold dots) and/or discharging by thermal electrons.

ESCA has found two essential applications in the polymer field, which will be discussed in Sects. 2.4 and 2.5:

– analytical studies, especially of polymer surfaces [3, 18, 19]
– studies on the electronic structure of polymers [4, 5]

2.2 The ESCA Spectrometer

The block diagram in Fig. 2.3 shows the most important parts of an ESCA spectrometer.

The X-ray sources most commonly used are Mg K_α (1,253.7 eV, 0.99 nm) and Al K_α (1,486.6 eV, 0.83 nm). It has recently been suggested [20] to use harder excitation such as Ti K_α (4,510 eV) in order to achieve a greater escape depth δ of the electrons detached.

The sample (film or powder) is kept under vacuum of about 10^{-5} Pa on a sample holder which preferably allows for heating and cooling. The kinetic energy of the elec-

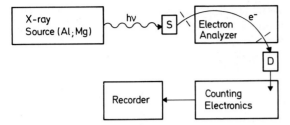

Fig. 2.3. Block diagram of a typical ESCA spectrometer; explanation see text

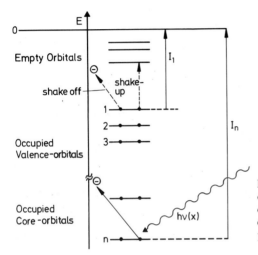

Fig. 2.4. The origin of shake-up and shake-off satellite peaks in core level spectra. The dotted arrows indicate examples of processes coupled with the main ionisation process involving a core orbital (n)

trons detached is determined in most commercial instruments using a double-focussing electrostatic analyser. The number of electrons of a given energy is counted and recorded as a function of their kinetic energy.

2.3 Main Features of ESCA Spectra

As pointed out in Sect. 2.1, the ESCA spectrum consists of a plot of the (relative) number of electrons as a function of their kinetic energy or, transformed using Eq. (2.1), of a related plot as a function of binding (= ionisation) energies. The electrons detached and analysed either originate from loosely bound outer ("valence") electrons or from tightly bound inner ("core") electrons. If a X-ray photon removes an electron from the n-th (core) orbital, the energy I_n is consumed and the corresponding peak in the spectrum is in general strong. With a certain probability, simultaneous energy-consuming processes may occur, as indicated in Fig. 2.4, which decrease the kinetic energy of the detached core electron and cause satellite peaks at lower E_{kin}. The energy differences to the corresponding main peak are similar to those observed in ionisation (shake-off) or optical excitation (shake-up) of valence electrons, although the selection rules are different in the latter case.

9

B. Electronic Spectroscopy

Shake-up and shake-off occur simultaneously with the ejection of a core electron. Other processes follow after these processes, since the electron system is left in a highly excited and short-lived state which is deactivated by
– X-ray fluorescence (emission of photons)
– electron emission (Auger effect).

The Auger electrons are also recorded and can be distinguished from true "ESCA electrons" using different excitation wavelengths the kinetic energy of Auger electrons being independent of the exciting photon energy.

The information gained from the spectral position of ESCA peaks relates to the binding energies of core and valence electrons and to energy differences between the different orbitals. The line-width gives further information, provided the influence of instrumental broadening can be minimised. The total line-width (Δ) of core-electron peaks is composed of three additive terms, Eq. (2.2)[3]. Related to the half-width of X-ray excitation Δ_x, the half-width due to the spectrometer (electron analyser), Δ_s, and the natural line-width Δ_r.

$$\Delta^2 = \Delta_x^2 + \Delta_s^2 + \Delta_r^2. \tag{2.2}$$

For non-monochromatised Mg or Al(K_α) radiation, Δ_x amounts to 0.7 and 0.9 eV, respectively, for monochromatic Al radiation $\Delta_x = 0.55$ eV has been reported[4]. In modern spectrometers, Δ_s can be neglected. The natural-line width is related to the uncertainty principle, Eq. (2.3).

$$\Delta E \cdot \Delta t \geq h/2\pi \tag{2.3}$$

and thus determined by the lifetime of the excited positive hole left by electron detachment, which is in the range between $\tau = 10^{-17}$ to 10^{-13} s. According to Eq. (2.3), with $\tau = \Delta t$ for $\Delta_r \,(= \Delta E)$ we obtain a range of 10^{-2} to 100 eV. Atoms with medium atomic masses have natural line-widths of the order of $\Delta_r = 0.1$ to 1 eV and thus lie in the region of Δ_x. The very broad lines of the core electrons of higher elements ($\Delta_r \sim Z^2$) are of little analytical use and not very interesting if polymers are considered.

Core electron peaks are prominent in ESCA spectra since their binding energy is nearer to the photon energy of exciting X-rays than that of weakly bound valence electrons thus facilitating the energy uptake. Core electrons are less directly involved in binding the atoms to monomers and polymers, compared to the valence electrons which belong to molecular rather than to atomic orbitals or to energy bands (Sect. 2.5). Core electrons are, however, influenced by the mode of binding, as evidenced by shifts in ionisation energies relative to the free atom values. The free atom values are characteristic of the element and the orbital considered.

The shifts are called, by analogy to NMR, "Chemical Shifts", and depend on the electron densities near the nuclei on account of the shielding of the positive charge of the nuclei.

The core electron peaks in the ESCA spectra therefore can be used in order to identify the atoms present in the polymer – and more specifically near the surface –, their mode of binding and their relative abundance.

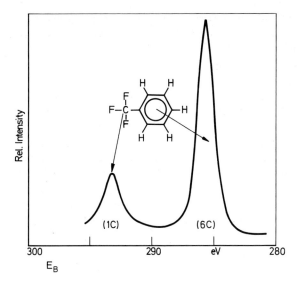

Fig. 2.5. ESCA spectrum of benzotrifluoride; after Clark [3]

2.4 Applications of Core-Electron Spectra

2.4.1 The Chemical Shift

The spectrum of the core electrons is most frequently used for analytical applications, although the valence electrons also contain analytically important information. The reason for this preference is the high intensity of the core peaks which is due to the high ionisation cross section of core electrons for X-rays. As a rule, ionisation is the more efficient the better the excitation energy matches the ionisation energy, i.e. if nearly the total X-ray photon energy is used for detaching the electron and the kinetic energy of the electron left over according to Eq. (2.1) is small. Valence electrons, on the other hand, use only a small portion of the exciting photon energy and therefore are ionised only with smaller probabilities.

The core electrons (e.g. C $1s$, N $1s$ etc.) are primarily localised at the individual atoms that form the polymer and therefore are not considered to contribute to chemical binding. They are, however, influenced by the "chemical environment" within the molecule so that their ionisation energies somewhat vary, depending on the nature and number of neighbouring atoms. The most marked effect is exerted by electro-negative partners which decrease the shielding of the positive nuclear charges by outer electrons, thus increasing the effective attractive force of the nucleus with regard to the core electrons. The ESCA signals of core electrons give the following analytically important information:

a) identification of the elements contained in the polymer
b) mode of binding (in favourable cases)
c) relative number of differently bound atoms (if chemical shift $> \Delta$).

A pronounced example of chemical shifting of the C $1s$ signal is given in Fig. 2.5. ($C_6H_5CF_3$, benzotrifluoride), showing that the binding energy of C $1s$ electrons in the CF_3 group is 8 eV higher than in the phenyl groups. The relative number of the two species of C atoms is reflected in the peak area ratio. The magnitude of the shift depends

11

B. Electronic Spectroscopy

Table 2.1. ESCA chemical shifts of several vinyl polymers, relative to PE

Polymer (Abbreviation)	R	Chem. shift of α C $1s$[a] eV
Poly(ethylene) (PE)	—H	0.0
Poly(vinylfluoride) (PVF)	—F	3.1
Poly(vinylchloride) (PVC)	—Cl	1.8
Poly(vinylalcohol) (PVOH)	—OH	1.9
Poly(propylene) (PP)	—CH_3	-0.1
Poly(1-butene) (PB)	—C_2H_5	-0.2
Poly(styrene) (PS)	—C_6H_5	-0.6

[a] Positive sign means increase a binding (ionisation) energy

on the nature and number of substituents. Table 2.1 gives some examples of α C $1s$ in different vinyl-type polymers.

$$\left[-CH_2-\underset{\underset{R}{|}}{\overset{\overset{H}{|}}{\underset{\beta}{C}}}- \right]_n$$

Since the natural ESCA line-width of the light atoms considered and the exciting X-ray line-width are each about 0.5 eV, only the larger, positive chemical shifts, which are due to electro negative substituents can be used for analytical purposes. Comparing PVF (Table 2.1) with benzotrifluoride (Fig. 2.5) we see that each fluorine atom contributes an increment of about 3 eV to the total chemical shift and even the βC $1s$ peak in PVF is still shifted by $+0.8$ eV/F [3]. The high increment of the fluorine atom makes ESCA uniquely suited for studying fluorinated polymers, e.g. Teflon (PTFE), where many other methods fail; but also oxygen- and chlorine containing polymers can be analysed due to the high electro-negativity of the substituents (Table 2.1). As an example of oxygen-containing polymers, the ESCA spectrum of polyethylene terephthalate (PET) is shown in Fig. 2.6.

$$\left[-CH_2-O-\overset{\overset{O}{\|}}{C}-\bigcirc-\overset{\overset{O}{\|}}{C}-O-CH_2- \right]_n$$

PET

This polymer has three differently bound groups of C-atoms (aliphatic, ester, aromatic) in a ratio 2:2:6 and two differently bound groups of (ester) 0 atoms in the ratio 1:1.

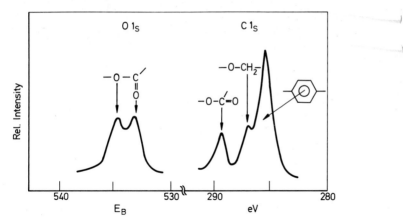

Fig. 2.6. ESCA spectrum of PET; after Clark[3]

The spectrum shows in the C 1s region the aromatic signal near 285 eV (cf. Fig. 2.5) and two peaks shifted to higher binding energies due to electro-negative substitution with one (287 eV) and two oxygens (289 eV). The oxygen increment therefore amounts to about 2 eV/O in PET, the same value as was found for PVOH (Table 2.1). The 0 1s-electron is bound more tightly, due to the higher effective nuclear charge of oxygen compared to carbon. The two differently bound 0 atoms again show distinct signals.

The magnitude of the chemical shifts can be well correlated with empirical electro-negativities[4] of the substituents.

2.4.2 Bulk and Surface Analysis

The molecules and atoms detected by ESCA always belong to the surface of the polymer, $\delta = 1$ to 2 nm being a good approximation of the $1/e$ escape depth of fast and slow electrons. The escape depth of medium energy electrons ($E_{kin} \approx 50$ to 100 eV) is smaller, $\delta \approx 0.5$ nm[3]. Depending on the sample preparation, the information gained by ESCA of polymers may be characteristic of the bulk of the solid polymer as well. In this case, each contamination of the surface, e.g. oxidation, must be scrupulously avoided. The purity of the surface can be tested by comparing peaks originating from different depths due to the different kinetic energy of the electrons released.

In surface studies, on the other hand, the specific treatment to be investigated – e.g. Corona discharge, chemical modification, photo-oxidation – is applied to the polymer prior to the spectroscopic investigation. As an example of surface modification, the pressing of PTFE powder between Al foils should be mentioned[3]. At the lowest pressing temperature possible (200 °C), only typical C 1s, F 1s, and F 2s peaks occur together with Auger electrons which can be distinguished by changing the excitation source. At a pressing temperature of 300 °C, additional peaks occur which are typical of O 1s, Al 2s, Al 2s, and Al 2p point to a very thin (<1 nm) layer of Al_2O_3 formed at the surface of the PTFE. This very thin layer cannot be identified by any other spectroscopic method.

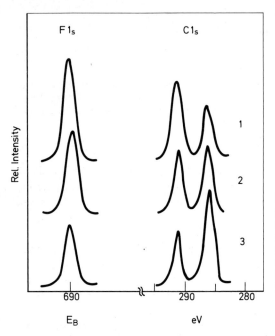

Fig. 2.7. ESCA spectra of PTFE/E copolymers: 63 mol% C_2F_4 (curve 1), 44 mol% C_2F_4 (curve 2) and 32 mol% C_2F_4 (curve 3) after Clark [3]. Composition calculated from $(A+B)/C$. Predicted composition from monomer reactivity ratio: 63 mol% (1), 50 mol% (2) and 36 mol% [3]

2.4.3 Examples of Bulk Analysis

Under clean surface conditions, the ESCA spectrum is representative of the bulk of a polymer sample, e.g. Fig. 2.6. The data presented in Table 2.1 can be used to identify these simple vinyl polymers. Structural details, as head-to-head links, branching, residual double bonds and conformation, are in general not indicated in the ESCA spectra.

The analysis of copolymers is an important application of ESCA, provided the core levels of the monomers are well separated. Systems which have already been investigated mostly comprise fluoro polymers [3], e.g. the copolymers of tetrafluoroethylene and ethylene [6] shown in Fig. 2.7. The evaluation can be based upon the ratio A/B, using C $1s$ peaks only (no calibration needed), or on the ratio $(A+B)/C$. In the latter case, due to the much different energy of the F $1s$(C) electrons, a calibration using pure homo polymers has to be performed.

Recently, the in situ synthesis of polymers using plasma techniques has become an accepted method. These polymers are highly cross-linked and insoluble so that ESCA turns out to be a very useful tool for characterisation [19, 20] of these polymers.

2.4.4 Examples of Surface Analysis

Surface reactions are conveniently studied by ESCA, as already demonstrated by the example of PTFE heated between Al foils [3, 19, 20]. Another example regarding this polymer, which has a low adhesive strength due to its low surface energy, is given by the action of sodium metal, dissolved in liquid ammonia, in order to chemically modify the polymer surface. The ESCA spectrum of PTFE etched with sodium solution does not contain any fluorine signal; this indicates complete chemical reaction in a layer exceeding 2 nm in depth. A polymer sample treated in this way rapidly takes up oxygen from the air, which is bound in groups like CO, COOH, etc.

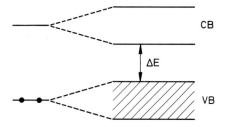

Fig. 2.8. Schematic view of energy band and energy gap formation in a crystal consisting of many atoms. The lower atomic level is filled with electrons, the upper one is empty

Other treatments of the surface of different polymers to be studied by ESCA include:
- corona discharge (formation of highly reactive radicals, atoms, excited molecules, etc. under the action of electrons emitted from a wire kept under high electrical tension), see also Chap. 7,
- ozone treatment and photo-oxidation,
- CASING (Cross-linking by Activated Species of Inert Gases)[3]. In this method, accelerated ions (e.g. Ar) are ejected to the surface under exclusion of oxygen. The primary radicals recombine under cross-linking and thus improve the mechanical properties of the surface. In PTFE/E, the CF_2/CF ratio decreases after very short treatment; this indicates the removal of F atoms from the surface,
- plasma treatment[7]: a microwave discharge at low pressure creates fast electrons, ions and excited particles; UV/VIS radiation originates from the deactivation of the latter. As shown by ESCA, the effect of this treatment on PTFE/E is similar to that of CASING.

2.5 ESCA Spectrum of Valence Electrons

2.5.1 Energy Bands

The term "Energy Band", as developed in solid state physics[8], visualises the fact that in most solids (metals, semiconductors, insulators) the outer electrons of the lattice constituents do not occupy well-defined narrow energy levels, as in free atoms, but rather a dense band of closely spaced levels. The energy bands are separated from each other by forbidden zones which are called energy gaps. In free atoms we can distinguish (fully or partly) occupied and empty levels or orbitals. The same is true of energy bands in solids; the occupied bands are called valence bands (VB), the empty ones conduction bands (CB).

The theoretical models describing the formation of energy bands are connected with the names of Bloch and Brillouin[9]. According to Bloch, first an atom is brought into contact with a second one; this results in a splitting of each energy level into a bonding and an anti-bonding one, the magnitude of splitting (in energy units) being twice the exchange interaction energy (J_A). This simple case describes the formation of a molecule consisting of two atoms. The splitting of core electron levels is insignificant.

The formation of a crystal leads to
1. interaction of many atoms instead of two
2. ordering of the atoms according to crystal symmetry and lattice constants.

The first effect induces further splitting of individual levels until energy bands are formed, as indicated schematically in Fig. 2.8. Using another picture, the valence elec-

15

B. Electronic Spectroscopy

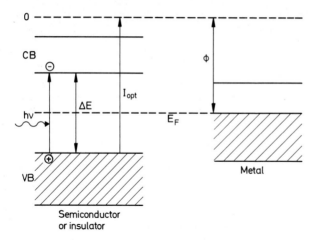

Fig. 2.9. Simple band model of a pure semiconductor or insulator (e.g. a polymer) in equilibrium with a metal of equal work function. As usual, the bands are sketched using a non-specified space coordinate

trons are tunnelling through the potential barriers which separate the atoms of the lattice. The width of the VB then gives the probability of this electron exchange.

Often only the highest valence band and the lowest conduction band are considered, since these are responsible for most electric properties of a crystal. These two bands are separated by the energy gap ΔE whose magnitude distinguishes semiconductors ($\Delta E \lesssim 3$) from insulators ($\Delta E \gtrsim 3$ eV).

The second effect (lattice ordering and symmetry) implies that the interaction leading to energy bands is periodic and anisotropic, as required by the lattice. This is immediately evident for every chemist, especially if the highly directional covalent bonds are considered. In solid state physics, this fact is expressed by the "reciprocal lattice", a useful concept for interpreting diffraction phenomena[1] which replaces lattive vectors[8] by their reciprocals and allows the energy of electrons in the bands to be calculated as a function of their "wave number" (k) in the reciprocal lattice. The close analogy between diffraction and energy band formation is stressed in Brillouin's approach, starting from free electrons in a lattice of positive atomic truncs, and allowing for interference of the electron waves according to de Broglie. This aspect of the band model will not be discussed in detail, since the $E(k)$ presentation of energy bands in polymers has only been used in theoretical calculations and was replaced by the density of states $D(E)$ presentation in recent papers. The density of states is defined as the number of energy levels per energy interval and is proportional to the ESCA signal, provided the ionisation cross section (efficiency) is constant over the energy region considered.

Before discussing the application of the band model to polymers, some basic notions have to be explained, using a simple model consisting of highest (filled) valence band and lowest (empty) conduction band (Fig. 2.9). Irradition of a semiconductor or insulator with light of photon energy $hv \geq \Delta E$ creates charge carriers – electrons in CB, posi-

---

1 The diffraction pattern, e.g. of X-rays, caused by a lattice provides a picture of the reciprocal lattice, the real lattice has to be calculated using Bragg's formula

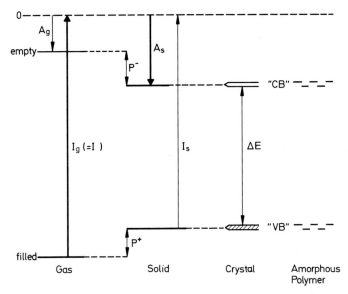

Fig. 2.10. Schematic view of energy levels in the free molecule and in crystalline and amorphous solids. This case is relevant to polymers showing side groups with lower ionisation energies and/or higher electron affinities compared to those of the backbone

tive holes in VB – moving in opposite directions if an electric field is applied or vanishing by recombination. By increasing the frequency of the exciting radiation we reach I_s, the ionisation energy of the solid, as already discussed in Sect. 2.1. In the case of metals, which are characterised by a partly filled valence band, the corresponding energy is called work function (Φ) and is, here, identical with the Fermi level (E_F) or chemical potential of the electrons. In ideal (pure) semiconductors and insulators this level is located at $\Delta E/2$ between VB and CB and does not correspond to any real, i.e. allowed electronic energy level. This term is important when fixing the zero-level in polymer ESCA experiments (Sect. 2.5.3).

2.5.2 Energy Bands in Polymers

The width of energy bands reflects the strength of interaction between the lattice constituents. Two extreme cases can be distinguished:
a) covalent crystals, e.g. diamond (insulator) or silicon (semiconductor) showing strong interaction and correspondingly broad bands
b) molecular crystals, which are typically held together by weak van der Waals forces and thus display only very narrow bands.

Case b) is characterised by the fact that the lattice units are modified by their surroundings, but have not lost their molecular identity or "individuality" in contrast to the atoms forming a covalent crystal (a). The modification of energy levels in b) is small with regard to optical transitions (Chap. 3), but significant with regard to the energy levels leading to VB and CB, as shown in Fig. 2.10. Starting with the free molecule or polymer side group, we see that the first ionisation energy is decreased and the electron affinity increased by polarisation of the solid environment [10, 22, 23]:

$$A_s = A_g + P^- \qquad P^+ \approx P^- \approx P$$
$$\underline{I_s = I_g - P^+}$$

$$\Delta E = I_s - A_s \approx I_g - A_g - 2P. \tag{2.4}^1$$

Considering the molecules or polymer side groups as belonging to a crystal lattice, we assume splitting to narrow energy bands which are lost again if the crystal lattice disappears during melting or formation of an amorphous solid: in this case, each molecule or side group "feels" a somewhat different microscopic environment with regard to distance and – more importantly [11] – with regard to the relative orientation of the individual molecules. The breakdown of the band model in weakly interacting amorphous solids leads to isolated, albeit closely spaced (<0.1 eV) energy levels. At this point we surpass the limits of ESCA, since this method is presently not yet able to detect these subtle energy differences. However, ESCA is highly relevant with regard to the energy bands of the polymer backbone, showing strong interaction along the covalent bonds (broad intramolecular energy bands), but only weak van der Waals interactions perpendicular to the chain direction.

Polymer chains therefore can be treated as pseudo- (one-dimensional) crystals; model calculations based on this assumption reproduce the sequence of ESCA valence peaks satisfactorily [12, 4, 5].

2.5.3 The Zero-Level Problem in ESCA Studies of Valence Bands

Although the valence signals in ESCA experiments are in general weak, they can be improved by cumulation of spectra and averaging in order to improve the S/N ratio. The main experimental problem consists in determining the zero energy level of the electrons detached, which is altered by not entirely avoidable static charging effects. Usually, a very thin layer of gold is evaporated onto the polymer film and thermal equilibrium between the metal and the polymer is assumed, i.e.

$$E_F \text{(polymer)} = E_F \text{(Au)}.$$

In these experiments, the ESCA spectrum consists of peaks characteristic of the polymer and the metal, the Fermi level being taken as the common zero level. If it is true that the Fermi level of the polymer is situated exactly in between VB and CB, ΔE can be calculated from the ESCA spectrum. In order to calculation the absolute energy of the levels, either A_s or I_s has to be known from independent measurements. For polyolefins the solid state electron affinity is supposed to be $A_s \approx 0$. Even without exact calibration, the ESCA spectrum yields the relative energies and the width of the valence bands which is useful for comparison with theory.

2.5.4 Examples of Valence Band Studies Using ESCA

Many linear polymers consist only of a backbone (HDPE, POM) or have side groups which are characterised by tightly bound valence electrons (PP, PB, LDPE). These

---

1 A_g is negative if the electron is not bound to the molecule in an exothermic reaction; $A_g \approx 2$ to 3 eV in typical electron acceptors, $I_g \approx 8$ to 12 eV in the average case, 6.5 to 8 eV in pronounced electron donors. $P \approx 1$ to 2 eV in most cases (molecular crystals) studied

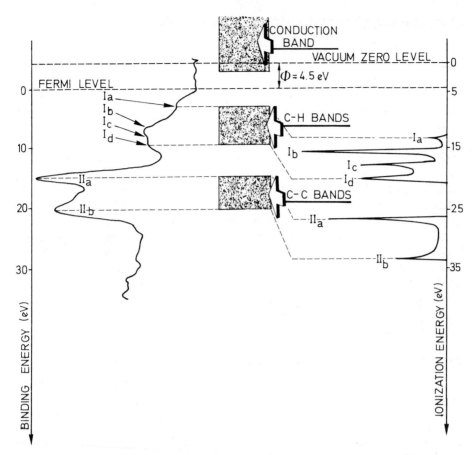

Fig. 2.11. Band structure of PE, calculated and measured by ESCA; after André et al. [13]

polymers can be treated theoretically using the "Extended Hückel" method or by "ab initio" calculations in order to calculate the valence bands [4]. In these calculations, the polymer chain is usually treated as an infinitely long, extended chain, neglecting more complicated conformations and intermolecular interactions. The energy bands calculated in this way are one-dimensional or intra- (macro-) molecular bands.

Exact coincidence of calculated and experimental band energies therefore cannot be expected and has not been observed. However, the relative position and internal structure of the bands $D(E)$ is well reproduced by more recent calculations [4, 13], as can be seen in Fig. 2.11. The structure in the bands results from the non-uniform distribution (density) of energy levels within the bands whose width typically amounts to several eV.

As an example, linear polyethylene (HDPE) will be discussed first [12]. It should be noted that the linear paraffin $C_{36}H_{74}$, which can be considered as an oligomer of HDPE ($N = 18$) behaves quite similar to the high-polymer with respect to its band structure. The ESCA valence region of HDPE consists of three regions [13], see Fig. 2.11:
1. a weakly structured band 4 eV below E_F, corresponding to the highest valence band, originates from C—H bonds,

B. Electronic Spectroscopy

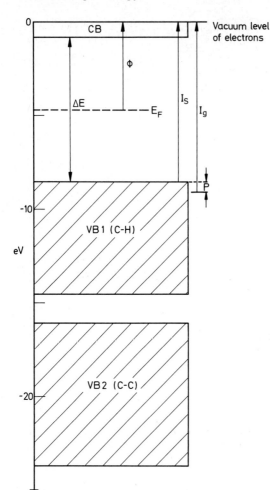

Fig. 2.12. Schematic energy band diagram of PE, based on ESCA and photoionisation experiments (see text); Φ can be estimated from electrostatic charging experiments [15]

2. the C—C (C 2s) valence band shows higher binding energies and is divided into two sub-bands due to "bonding" (18.8 eV) and "antibonding" (13.2 eV) levels, both fully occupied by electrons. The splitting of this band is well predicted by theory,
3. the third region is due to inelastically scattered electrons detached from both bands.

The analysis of the spectrum reveals the following, still preliminary band structure of HDPE (Fig. 2.12): about 4 eV below E_F, the reference level of the experiment, there is the first (C—H) valence band whose four maxima of $D(E)$ can be recognised in the spectrum as a weak structure. If the assumption of E_F being situated exactly at $\Delta E/2$ is correct, we obtain

$$\Delta E\,(\text{HDPE}) = 7.7 \pm 0.5 \text{ eV}.$$

The second VB, which is about 8 eV broad, is separated by a narrow forbidden zone (1.6 eV) from the first one.

The absolute energies of the bands are still controversial [14]. Photoemission of electrons starts at $I_s = 8.5$ eV in good agreement with $E_F = 4.5$ to 5 eV, (estimated by other methods), since $E_F = I_s - \Delta E/2 \approx 4.65$.

The lower edge of CB (A_s) which is not measurable by ESCA seems therefore to be situated near the vacuum zero level of the electrons ($A_s = +0.5$ to -1 eV). A negative value A_s would mean that there are no states available for excess electrons in pure PE. The ionisation energy of long paraffin chains can be extrapolated to $I_g = 9.0$ eV, giving $P = I_g - I_s = 0.5$ eV. This low value of P seems reasonable, considering charge spreading in large molecules.

The ESCA valence spectrum of PVC, taken as a second example, can be explained as a superposition of the PE spectrum and structures due to chlorine (Cl $3p$ – "lone pairs" – at 5.7 eV; Cl $3s$ at 17.5 eV, relative to E_F). The chlorine levels are more atomic and less band-like in character compared to the C—C band which can clearly be identified on both sides of Cl $3s$. Making the same assumptions as above, we obtain $\Delta E = 9.8$ eV and $A_s \approx 0$.

Data on valence bands have been obtained for about ten polymers, ESCA being the only method available for these studies. The absolute values are still uncertain by about 1 eV, even though the most elaborate technique known at present has been used. The calculations are uncertain in this respect by about 10 eV but nevertheless offer invaluable help in identifying the bands and their sub-levels.

Beyond the purely scientific interest, these studies contribute very significantly to our understanding of the electronic structure of polymers and thus of [15, 16]
– photo conductivity
– electrostatic charging and
– adhesion
of polymers, the latter two items being typical surface phenomena and thus particularly suited for ESCA studies. Localised levels often will have to be considered in addition to band levels, especially in polymers with large ΔE.

2.6 Résumé of ESCA in Polymer Spectroscopy

The application of ESCA to polymers is a recent development. It is the only method which allows both valence- and core electrons to be measured. The results are the energies of the orbitals or bands provided, as tacitly assumed in the discussions above, that Koopmans theorem is valid. This condition is fulfilled except that very fast relaxation processes occur during ionisation, i.e. if the electrons ejected to not change their orbits during the elementary processes. This assumption seems to be valid within about 10 percent. Thus, ESCA measures approximately the energy of the electrons in their unperturbed orbits, the actually measured value being the difference between the initial and the final state of the polymer in this experiment.

Table 2.2 shows the matrix presented in Sect. 1.2, as applied to ESCA. As can be seen, the application of ESCA has so far been restricted to analytical work and electronic structures. Conformation should have an influence on 1D valence bands. Complex formation could influence the valence electron spectra or possibly appear in the shake-up structure of the core electron spectra.

The greatest potential for applications of ESCA in polymer research seems to be the investigation of

B. Electronic Spectroscopy

Table 2.2. Information obtained by ESCA relating to structure and dynamics of polymeric systems

Structure		Dynamics	
Chemical structure	+	Movements of the chain,	
Tacticity	−	segments and side groups	−
Conformation	−	Phonons	−
Crystallinity	−	Excitons	−
Electronic structure	+	Complex formation and related phenomena	−

- insoluble polymers containing heteroatoms
- fluoro-polymers
- "plasma polymers"
- surface modification of polymers and
- valence bands and detailed electronic structure, including
- surface states of pure and contaminated polymers.

References

1. Siegbahn, K. et al.: Nova Scota R. Soc. Sci., Uppsala, Ser. IV, 20 (1967)
2. Siegbahn, K., in: Electron Spectroscopy. Progress in Research and Applications, Caudano, R., Verbist J. (eds.), Amsterdam: Elsevier 1974, p. 3; see also[17]
3. Clark, D.T., in: Jvin, K.J. (ed.): Structural Studies of Macromolecules by Spectroscopic Methods, London: Wiley 1976, p. 111
4. Pireaux, J.J., Riga, J., Verbist, J.J., Delhalle, J., Delhalle, S., André, J.M., Gobillon, Y.: Phys. Scripta *16*, 329 (1977)
5. André, J.-M., Ladik, J., Delhalle, J. (eds): Electronic Structure of Polymers and Molecular Crystals, New York: Plenum 1975
6. Clark, D.T., Kilcast, D., Feast, W.J., Musgrave, W.K.R.: J. Pol. Sci. Pol. Chem. Ed. *11*, 389 (1973)
7. Lee, L.-H. (ed.): Characterization of Metal and Polymer Surfaces, Vol. 2, Polymer Surfaces, New York: Academic Press 1977, Part 1 (ESCA)
8. Kittel, C.: Introduction to Solid State Physics, New York: Wiley 1971, German Transl. München: Oldenburg, Frankfurt am Main: Wiley, 3rd ed. 1973
9. Spenke, E.: Elektronische Halbleiter, 2nd ed., Berlin, Heidelberg, New York: Springer 1965
10. Gutmann, F., Lyons, L.E.: Organic Semiconductors, New York: Wiley 1967
11. Bässler, H.: Phys. Stat. Sol. (b), *107*, 9 (1981)
12. Delhalle, J., André, J.-M., Delhalle, S., Pireaux, J.J., Caudano, R., Verbist, J.J.: J. Chem. Phys., *60*, 595 (1974)
13. André, J.-M., Delhalle, J., Delhalle, S., Caudano, R., Pireaux, J.J., Verbist, J.J.: Chem. Phys. Lett. *23*, 206 (1973)
14. Bloor, D.: Chem. Phys. Lett., *40*, 323 (1976)
15. Bauser, H., Klöpffer, W., Rabenhorst, H. in: Advances in Static Electricity, Vol. 1, p. 2 (1970), Proceedings of the 1st Int. Conf. on Static Electricity, Vienna (Austria), May 4–6, 1970
16. Mort, J., Pfister, G.: Electronic Properties of Polymers, New York: Wiley 1982
17. Brundle, C.R., Baker, A.D. (eds.): Electron Spectroscopy, Theory, Techniques and Applications, Vol. 1–3, London: Academic Press 1977, 1978, 1979
18. Clark, D.T. in: Cantow, H.J. et al. (eds.): Molecular Properties, Adv. Pol. Sci., Vol. 24, 125, Berlin, Heidelberg, New York: Springer 1977
19. Clark, D.T.: Pure Appl. Chem. *54*, 415 (1982)

20. Clark, D.T.: ACS Meeting Monograph (1982/1983), The Modification, Degradation, and Synthesis of Polymer Surfaces Studied by Means of ESCA
21. Peeling, J., Clark, D.T.: Pol. Degrad. Stab. *3*, 97 1980
22. Silinsh, E.A.: Organic Molecular Crystals, Their Electronic States, Berlin, Heidelberg, New York: Springer 1980
23. Pope, M., Swenberg, C.E.: Electronic Processes in Organic Crystals, Oxford University Press 1982
24. Seki, K., Hashimoto, S., Sato, N., Harada, Y., Ishii, K., Inokuchi, H., Kaube, J.: J. Chem. Phys. *66*, 3644 (1977)

3 Absorption Spectroscopy in the Ultraviolet and Visible Regions

3.1 The far Ultraviolet (fUV)

Experimentally, the fUV ($\lambda < 200$ nm) is more demanding than the nUV/VIS-region to be discussed in the next sections, since the beam has to be conducted in vacuum and since the laboratory radiation sources are weak. Only very few solvents are transparent in the fUV. Polymers therefore have to be measured in the form of thin films and mostly show intense, structureless absorption bands (1–4) which become narrower and partly structured at low temperature. Qualitatively, these spectra resemble in the long-wavelength region of the fUV (about 100 to 200 nm) to absorption edges in insulators which are due to transitions from the energetically highest VB to the lowest CB. Unfortunately, the interpretation of the spectra is not always straightforward because of the pseudo-1D-band nature of polymers, i.e. their intermediate position with regard to electronic structure between true (3D) solids and molecules (see also Figs. 2.9., 2.10, and 2.12).

An experimental set-up for measuring fUV absorption spectra of thin polymer films has been described by George et al. [4, 5]. The radiation source is a micro-wave-(2.45 GHz)-excited H_2-discharge lamp yielding a weak fUV continuum at a pressure of about 1 mbar. In order to prevent intensity losses due to window absorption, the lamp is connected by a special slit with the grating monochromator, pumped to about 10^{-4} Pa. The resolution obtained with the 1 m grating monochromator fitted with 50 μm slits amounts to 0.15 nm. The polymer forms a thin (about 50 nm) film produced by tip-coating from dilute solution on to LiF discs [4] and is mounted in a cryostat if low temperature spectra are to be measured. There is again no window between the monochromator and the sample. The detection is effected by a photomultiplier behind sodium salicylate transforming the fUV into blue radiation. The apparatus described is a single beam spectrometer so that a second spectrum without sample has to be measured separately in the same arrangement. The signal of the detector has to be corrected for sample fluorescence if the polymer emits radiation. As an example of the few fUV-polymer spectra reported, the spectrum of PE will be discussed according to the results obtained by Partridge [1] and George et al. [4]. It should be noted that hardly any differences exist between the spectra of LDPE, HDPE and long-chain paraffins ($C_{28}H_{58}$ and $C_{42}H_{86}$), i.e. branching and chain length seem not to be important in polyolefins if $n \geqslant 14$.

Figure 3.1 shows the "absorption edge" of PE at room temperature and at 4K. The absorption starts at about 8 eV (155 nm), the apparent absorption at longer wavelength

B. Electronic Spectroscopy

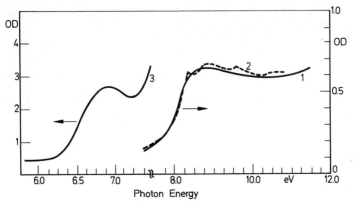

Fig. 3.1. fUV Absorption spectrum of HDPE at room temperature (curves 1 and 3) and at 4K (curve 2); film thickness roughly 50 nm (curves 1 and 2) and 40 µm (curve 3); after George [4]

being due to scattering. There is a broad maximum near 9 eV at room temperature and a second increase at about 11 eV. In the 4K spectrum, a sharper peak is observed at 8.35 eV (149 nm), and the absorption edge is somewhat sharper compared to that at room temperature. The asorption coefficient in the absorption maximum ($\alpha \approx 5 \times 10^5$ cm^{-1}) indicates very strong absorption, comparable to allowed band-to-band transitions of inorganic solids. Using thick films of PE (d ≈ 50 µm), a weak absorption band is detected at 6.90 eV (180 nm), which is absent in the spectra of oligomeric model compounds. This band is due to approximately one double bond per macromolecule, most probably to vinylidene end groups, as shown by independent IR evidence.

There are no π-electrons in pure PE; the strong absorption observed below 155 nm, therefore, can only be attributed to σ electrons. As shown in Sect. 2.5.4, photoionisation starts at 8.5 eV (146 nm) and may be responsible, together with photodissociations for the strong increase in absorption near 11 eV. The origin of the absorption edge is still controversial: is it due to absorption into an exciton state? Or do we observe a genuine VB→CB transition?

The difference between the two possible explanations is basically the attraction between electron and hole; if the Coulomb energy is higher than kT, electron and hole are coupled and form a shortlived neutral pseudo particle, the exciton; if it is smaller than kT, the particles move independently and carry the (intrinsic) photocurrent, provided an electric field is applied. Only in this latter case, therefore,

$$\Delta E = E_{opt} = E_{photo} \qquad (3.1)$$

do band gap, absorption edge and threshold energy of intrinsic photo current coincide (3.1). In PE, $E_{photo} = 8.81 \pm 0.05$ eV [6] and is 0.46 eV higher than the 4K absorption peak of PE (Fig. 3.1). This indication contradicting the VB→CB assignment is, however, not absolutely convincing, since photoconductivity being a macroscopic phenomenon, the charge carriers have to move between the polymer molecules so that the simple model of 1D energy bands may be insufficient to describe this process in polymers. A more serious argument against independent holes and electrons in PE, they may move in 1D or 3D bands, involves the low dielectric constant ($\varepsilon = 2.3$) causing mutual attraction of

24

Table 3.1. Examples of technical polymers containing nUV-chromophores

Polymer	Chromophore	Longest wavelength absorption nm
Poly(styrene)	Phenyl-group	270 [3, 16)], 280 (absorption edge) [35)]
Poly(ethylene terephthalate)	Terephthalate-group	290 (tail of absorption) [18, 46)] 300 [49)]
Poly(methyl methacrylate)	Aliphatic Ester	250–260 (absorption edge) [35)]
Poly(vinyl acetate)	Aliphatic Ester	210 (maximum) [9)]
Poly(vinyl carbazole)	Carbazole-group	345 [19)]

holes and electrons over distances up to 10 nm. The most likely explanation of the 4K absorption peak consequently involves a bound state (exciton) which may dissociate into a free charge carrier after additional supply of energy. The ESCA results (Sect. 2.5.4) are compatible with both interpretations.

The fUV-asorption spectra of very thin, oriented LDPE films, using the polarised radiation of a hydrogen lamp have been reported recently [32)]. According to this work, the absorption edge near 8 eV is observed if the polarisation of the electric field vector is perpendicular to the drawing direction and, hence, perpendicular to the main chain of PE. Synchrotron radiation reveals a first, "parallel" absorption edge near 11.5 eV [32)]. The dichroism observed agrees well with ab initio calculations [33)] although the calculated energy gap is much too high.

The measurements do not, however, resolve the problem of exciton vs band absorption.

Polystyrene [2, 3)] shows predominantly π-electron excitations, so that the absorption edge of the back-bone cannot be assigned accurately.

Solutions of Poly(oxyethylene) in water and Poly(vinyl acetate) in acetonitrile have recently been investigated [9)] down to 170 nm. In this region, commercial spectrometers can still be used.

At energies above 10 eV, photo-ionisation and numerous photo-dissociation processes occur which have not yet been investigated in polymers.

Due to the experimental difficulties, the fUV-spectroscopy of polymers is still in its infancy. The use of Synchrotron radiation [7, 8)] as a powerful source of fUV may remove some of the difficulties, so that at least the basic research in this field, complementary to ESCA and photoconductivity, may increase in importance in future.

The main fields of application (according to Table 1.1) are clearly "Electronic structure" and "Excitons". Formally, UVPES (Sect. 2.1) belongs to the fUV-region. This variant of ESCA seems appropriate to study valence electron spectra, since the excitation energy is nearer to the binding energies of these loosely bound electrons.

3.2 Absorption Spectroscopy in the Near Ultraviolet and Visible Regions

3.2.1 Spectral Region and Characteristic Transitions

The nUV and the VIS regions of the electromagnetic spectrum spectroscopically form one entity, since with a few exceptions we observe the absorption and emission (luminescence) of π- and n-electron systems in this region, often exemplified by intense transitions and characteristic vibrational structures. The nUV is that part of the UV which can be observed in air ($\lambda \geqslant 200$ nm) and is next to the shortest visible wavelengths ($\lambda \approx$ 400 nm). Next are the regions of the visible spectrum, giving rise to the visual sensations known as spectral colours. Polymers generally absorb in the VIS region only if purposely modified or degraded, the exceptions being highly conjugated polymers used in studies of electrical conductivity or catalytic activity. Biopolymers are often coloured by specific chromophoric groups, e.g. hemins or carotenoids. Only few dyes and complexes absorb electronically above a wavelength of about 750 nm, the red edge of the VIS region.

3.2.2 Basic Principles of nUV/VIS Absorption Spectroscopy[10, 11]

3.2.2.1 Electronic Vibronic Spectra

The absorption of photons in the nUV/VIS region by organic molecules is caused by excitation of π- and n-valence electrons including the additional excitation of molecular vibrations. Only the O—O transition (Fig. 3.2) occur between the vibrational ground states of the electronic ground state (mostly S_0) and electronically excited states (S_n). Potential curves can be depicted only for diatomic molecules[12], as shown in Fig. 3.2. The solid curved denote the return points in the classical (mechanical) picture of two vibrating atoms. The highest probability according to quantum mechanics, in the zero vibration mode is at R_0, the coordinate of the minimum of the classical curve; the first excited vibration has two maxima, etc. If the equilibrium bond length is equal in S_0 and S_1, the O—O absorption is the strongest one. The electronic states considered in absorption spectroscopy of polymers are mostly singlet states $S_0, S_1, S_2 - S_n$), all electron spins being compensated (symbolic: $\uparrow\downarrow$).

The graphic presentation using potential curves and discrete vibrational levels corresponds to the Born-Oppenheimer approximation in molecular quantum theory,

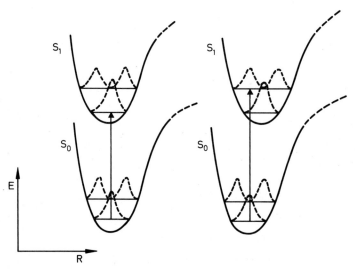

Fig. 3.2. Schematic presentation of a O-1 absorption with equal and unequal atomic coordinates in ground and excited state

based on the separation of the wave function [Eq. (3.2)] into electron and nuclear wave functions which only depend on the

$$\Psi(\mathbf{r}, \mathbf{R}) = \psi(\mathbf{r}) \times \varphi(\mathbf{R}) \tag{3.2}$$

electronic (**r**) and nuclear coordinates (**R**), respectively. The physical basis of this model rests on the vastly different velocities of electronic and nuclear movements so that the former can be calculated assuming that the latter is constant during the electron movement.

3.2.2.2 Energy Level Diagrams

UV-spectroscopic results are graphically presented using diagrams like the one shown in Fig. 3.2 or, more frequently, a Jablonski-Diagram (Fig. 4.1). The energy levels indicated always correspond to measured or calculated energy differences. Spectra therefore are best plotted to show absorption intensity (optical density or absorption coefficient) vs. wave number (v'), since this magnitude is proportional to energy (Fig. 1.1); in this way the spectrum itself is kind of an energy level diagram. Molecular orbital diagrams, as derived from simple MO models or using measured ionisation energies (Chapt. 2) and electron affinities, are of little value for quantitative comparison with UV spectra. The reason for the large discrepancies ($I_g - A_g \gg E_{opt}$) lies in the fact that the orbitals have no fixed energy, irrespective of the number of electrons occupying them. The electron lifted by absorption of a photon into a higher orbital strongly interacts with all other electrons and especially with the positive hole remaining in the originally filled MO. Thus, the energy differences measured in absorption spectra do not simply correspond to the differences in unperturbed orbital energies. Electronic spectra can be calculated using elaborate theoretical models involving electron/electron interactions and mixing of MOs [11].

3.2.2.3 Intensity of Absorption Bands

Complex chromophores in Polymers or their monomeric models always contain more quantum mechanically allowed electronic states than can be observed in optical absorption spectra due to the selection rules. The two most important selection rules are connected with the electron spin and with the symmetry of the MO's involved. The spin selection rule, according to Kasha [13] the strictest one, forbids the electron spin (Chap. 9) to be changed during absorption or emission of a photon.

Absorptions between states of different multiplicity, e.g. $S_0 \rightarrow T_1$, cannot be observed without tricks (oxygen pressure technique, heavy atom effect).

The symmetry selection rule strongly reduces the intensity of absorption if the positive nuclei and negative electrons have the same centre of gravity in the ground and excited state, i.e. if excitation does not lead to a transfer of charge. In this case, the transition dipole moment ($\mathbf{M}_{k,n}$) which measures the strength of an absorption band [Eq. (3.4)] equals zero. The transition dipole moments is an integral of the form (3.3)

$$\mathbf{M}_{k,n} = e \langle \Psi_k | \sum_i z_i \mathbf{r}_i | \Psi_n \rangle, \tag{3.3}$$

where e is the elementary charge, Ψ_k, Ψ_n are the wave functions of the initial state (k) and the final state (n), z_i is the charge of the ith particle (electron or nucleus) and $\mathbf{r}_i$ is the space vector of the ith particle

$$f_{k,n} \sim \int \varepsilon dv' \sim v' |\mathbf{M}_{k,n}|^2 . \tag{3.4}$$

The oscillator strength $f_{k,n}$ of the transition (a relict of older mechanical theories of light absorption) is proportional to the integral absorption band and to the square of the transition dipol moment (3.4). Very strong, i.e. fully allowed electronic bands are characterised by:

$$f_{k,n} \approx 1,$$

$$\varepsilon_{max} \approx 10^5 \text{ l mol}^{-1} \text{ cm}^{-1} (\sigma_{max} \approx 4 \times 10^{-16} \text{ cm}^2 = 4\text{Å}^2),$$

$$|\mathbf{M}| \approx 10 \text{ D}[1].$$

This transition dipole moment roughly corresponds to a shift of one elementary charge over 2 Å.

3.2.2.4 The Franck-Condon Principle

Vibrationally structured electronic absorption bands show one of two principal band shapes: either the O—O peak is the strongest one, or one of the following (O-1, O-2 etc.) transitions has the highest probability and thus the most intense peak [Eq. (3.4)]. The reason for this behaviour is readily apparent from Fig. 3.2 and depends on the nuclear coordinates in the ground and excited states. The Franck-Condon principle implies that the most probable transition is the one showing best overlap of wave functions in the ground and the excited state without change in nuclear coordinates. The physical basis is again the high velocity of electronic compard to nuclear movements.

The weak intensity of the O—O peak may, however, also be due to symmetry (e.g. in the lowest UV band of benzene).

3.2.3 Application of Lambert-Beer's Law to Polymers

Lambert-Beer's Law is fundamental for quantitative absorption spectroscopy in the UV/VIS and constitutes a special case of the general law of absorption of radiation [Eq. (3.5)] in homogeneous matter.

$$I = I_0 \cdot e^{-\alpha d} \tag{3.5}$$

Equation (3.5) implies that the decrease in light intensity only depends upon the intensity at depth d and on a constant (α) characteristic of the substance. In the case of polymer films, especially if the absorbing units are not known, the absorption coefficient α according to (3.5) can be used in order to characterise the absorption behaviour of the polymer. In solutions and polymer films with known absorbing units, α can be written as the product of concentration × specific absorption.coefficient, provided that the particles absorb independently of one another. This condition, which is the essence of Lambert-Beer's law, is fulfilled especially at low concentration and therefore problematic for polymers. However, if there is a constant, i.e. concentration-independent interaction between the chromophores, Lambert-Beer's law can be used in the concentration range of constant interaction.

---

1 1 Debye (D) = $^1/_3 \times 10^{-29}$ Cbm

Depending on the concentration units employed, a variety of specific absorption coefficients can be used, e.g. the absorption cross section σ [Eq. (3.6)] or the molar decadic absorption coefficient ε [Eq. (3.7)] most frequently used in chemical absorption spectroscopy if the molar mass of the absorbing species is known so that c is known as well.

$$\alpha = N \times \sigma \tag{3.6}$$

N: number of absorbing particles (cm^{-3})
σ: absorption cross section (cm^2).

$$\alpha = 2.303 \, c \times \varepsilon \tag{3.7}$$

identical with

$$I = I_0 \, 10^{-\varepsilon cd} = I_0 \, 10^{-OD} \tag{3.8}$$

ε: molar, decadic absorption coefficient $(1 \, mol^{-1} cm^{-1})$
c: concentration of absorbing units $(mol \, l^{-1})$
OD: optical density or absorbance $(\log I_0/I)$.

If the molar mass of the absorbing species is not known, the specific absorption coefficient (ε') can be based on c' $(g \, l^{-1})$ as defined in Eq. (3.9)

$$\alpha = 2.303 \, c' \times \varepsilon'. \tag{3.9}$$

Since the intensities of absorption bands may vary over several orders of magnitude, it is common practice to plot the logarithm of the absorption coefficient, preferentially $\log \varepsilon$, vs. v' or λ. The applicability of Lambert-Beer's law can in principle be checked by measuring different concentrations of the polymer to be studied. However, unlike monomers, polymers never form really dilute solutions with regard to the absorbing groups, which may be present as side groups (A) or as integral parts of the backbone (B).

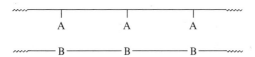

Polymers in which the chromophore extents over the whole molecule, e.g. polyconjugated molecules, are in most cases insoluble and should be treated by energy band models (Sect. 3.1) rather than as molecules.[1] Using Eq. (3.8), the molar mass of the basic unit (e.g. styrene for PS) has to be taken in order to calculate c and ε. The spectra obtained in this way can be compared directly with those of monomeric models (cumene or toluene in the example given) in order to detect deviations from purely additive behaviour (polymer vs monomer aspect, Sect. 1.2).

---

1 Insoluble polymers can be measured by reflection spectroscopy, the absorption spectrum being calculated[14]; results obtained in this way should be regarded with caution, especially if weak transitions are involved

The problem of chromophore interactions in polymer coils or networks is certainly smallest in the case of copolymers, especially if the UV-absorbing component is present only in low concentration; UV-absorption can therefore be used to determine copolymer composition.

In polymer films, the concentration of the chromophores is given by the chemical composition and the density of the solid polymer. In copolymers it may be varied by changing the monomer ratio. It is advisable also in this case to compare the copolymers with monomeric models the greatest experimental difficulties being the homogeneity of the films which usually have to be very thin.

Another experimental difficulty in evaluating UV/VIS-absorption spectra – particularly of partly crystalline films – consists in light scattering, adding to the true optical density (OD) an apparent absorbance [Eq. (3.10)] which is due to scattering [34].

$$OD_{obs} = OD + OD_{scat},$$ (3.10)

$$OD_{scat} \sim \nu'^n.$$ (3.11)

The additivity postulated by Eq. (3.10) is valid as long as the ratio of absorbing to non-absorbing groups does not exceed an upper limit which has to be determined experimentally ($\sim$ 1:15) in several detergents and proteins [15].

If (3.10) is valid, the Rayleigh Eq. (3.11) offers a simple possibility of correcting the spectra disturbed by scattering effects: outside the region of true absorption the spectrum has to be measured up to long wavelengths. Plotting log OD_{obs}, i.e. the measured absorbance, vs log ν' should give a streight line with slope n at long wavelengths. Extrapolating this line into the region of true absorbance and subtracting the OD's (rather than their logs!) yields OD (ν'), the true absorption spectrum. Care must be taken with impure polymers (e.g. technical PS grades) [16, 17], since small concentrations of impurities asorbing at long wavelengths may severely alter the spectral region useful for the scattering correction. On the other hand, such weak absorptions can be recognised if the spectrum has been recorded far into the non-absorbing range and the scattering correction has been applied.

3.2.4 UV-Absorbing Polymers

Well developed nUV absorption spectra can be obtained from polymers containing aromatic or heteroaromatic groups. Polymers showing $n\pi^*$ chromophores (C=O), are expected to give weak UV-absorption. Conjugated double or triple bonds give strong absorptions often detected in partly degraded polymers, e.g. PVC [18]. Some technically important polymers showing nUV-absorbing chromophores are summarised in Table 3.1. Collections of UV/VIS-spectra and reference books are quoted in the bibliography [36-43].

Among the biopolymers, proteins (due to tyrosine and tryptophane) and DNA/RNA (due to the purine and pyrimidine bases) strongly absorb in the nUV. With the exception of a generally weak hypochromy (Sect. 3.2.6) and some broadening of the spectra, the absorption spectra of pure polymers and their monomeric models are very similar, indicating the absence of any strong ground-state interaction between the chromophores. As examples for nUV-absorption spectra, PS and PVCA are shown in Fig. 3.3 in the log ε vs. ν' presentation.

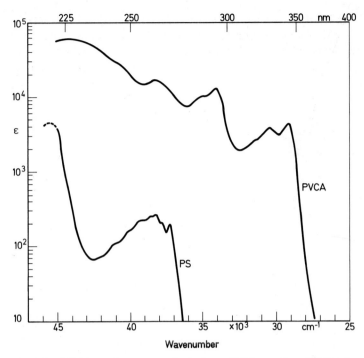

Fig. 3.3. nUV Absorption spectra of solutions of PS (curve 1) and PVCA (curve 2)

Table 3.2. nUV-Absorption of PVCA and NIPCA [19] oscillator strengths ($f_{o,n}$)

n	Absorption band (cm^{-1})	NIPCA in cyclohexane	NIPCA in methylene chloride	PVCA in methylene chloride	PVCA in solid state
1	29,000 (O—O peak)	0.012	0.010	0.012	0.015
2	34,000 (O—O peak)	0.050	0.045	0.049	0.058
3	38,000 (total)	0.17	0.16	0.17	0.18

Table 3.2 gives the wavelengths and molar decadic absorption coefficients of the O—O peaks of PVCA and, for comparison, N-Isopropyl carbazole (NIPCA). The considerable differences in ε_{max} between polymer and monomeric model are mostly compensated by band broadening in the polymer so that the oscillator strengths of the transitions f_{on} are roughly the same in the polymer and the monomer. Deviations will be discussed in Sect. 3.2.5.

The long wavelength band of PVCA ($S_0 \rightarrow S_1$) corresponds to the ^1L_b (Platt) or α (Clar) transition of monomeric carbazoles. This transition involves strong charge transfer from N to C 3 and C 6 and therefore is polarised in the short axis. The bands at 295 ($S_0 \rightarrow S_2$) and 261 nm ($S_0 \rightarrow S_3$) are polarised in the long molecular axis, the former corresponds to ^1L_a (Platt) or p (Clar) of carbazole and is due, simply speaking to the strong transition of an electron from the highest filled to the lowest empty MO.

The two band shown in the PS spectrum ($S_0 \rightarrow S_1$, $S_0 \rightarrow S_2$) are due to the weak, symmetry-forbidden ^1L_b transition shown by benzene itself and its simple alkyl derivatives and to the medium intensity ^1L_a.

The strong electronic absorption of crystalline polydiines [44, 45], e.g. TSHD (PTS) will be shown in connection with resonance Raman scattering (Sect. 6.6) and ESR of the growing chain (Sect. 9.7).

3.2.5 Hypochromy in Polymers

Deviations from the additive behaviour (monomer aspect) in UV-absorption of polymers can be caused by
a) shifting and splitting of bands
b) lowering of absorption intensity without spectral changes.

Changes according to (a) may occur in crystalline polymers (Davydov splitting [47]) not yet been definitely identified in polymers).

Hypochromy (b) has occasionally been observed in solutions of stereo-regular aromatic polymers [26]; the strongest effect, however, is displayed by DNA [20, 21]. UV-absorption of DNA is due to the four bases adenine, thymine, guanidine and cytosine, all strongly absorbing in the 200 to 280 nm region and yielding an absorption maximum near 260 nm ($f_{0,1} \approx 0.27$ in average base composition [21]). The total absorbance does not only depend on the base composition, but also on the degree of ordering. Homogeneous neutral solutions of denatured (random-coil) DNA in water show an absorption spectrum which can hardly be distinguished from that of a mixture of the bases and changes little upon hydrolysis to nucleotides. Native (helix form) DNA shows an absorption spectrum of the same form, but of 30 percent lower optical density over the whole absorption band. In order to understand this effect we have to remember than one turn of the double helix of DNA has a height of 3.4 nm and contains 10 base pairs, resulting in a vertical distance of 0.34 nm, the thickness of aromatic molecules. This close packing causes, among other effects, exciplex fluorescence [23] and hypochromy.

A semi-classical interpretation of hypochromy is based on the screening of neighbouring chromophores by dispersion interaction [23-26]. The UV-radiation induces dipoles in the chromophores which are felt by neighbouring chromophores as weakening of the oscillating electromagnetic field. This interaction leading to decreased absorption increases with the
– overlap of absorption spectra
– strength of absorption ($f_{k,n}$)
– decreasing distance of chromophores
– parallel or almost parallel arrangement of chromophores and thus their $\mathbf{M}_{k,n}$.

The spectral and structural data of DNA show that the first three conditions are nearly perfectly fulfilled, the fourth one is satisfied approximately. From the example of this biopolymer we may expect that strong transitions in synthetic polymers should give the same effect which thus could be used for identifying ordered structures when other methods fail, e.g. in dilute solution. It should be noted that the analysis of copolymers can be complicated by this and similar effects, especially if random and block copolymers are to be compared [28]. In order to avoid analytical errors it is imperative to carefully measure the spectra of the corresponding homopolymer(s) and a suitable monomeric model in addition to the spectra of the copolymers to be analysed. It should furthermore be remembered that the integrated absorption curve [Eq. (3.4)] rather than the maximum absorption coefficient is the true measure of absorption strength.

The importance of a careful error analysis in the UV-spectroscopy determination of copolymer composition has recently been emphasized by Garcia[47] in a critical evaluation of all data available on styrene polymers. He concludes that Lambert-Beers law is fulfilled within the limits of experimental error in all cases reported so far.

3.2.6 Absorption by Molecules dissolved in Solid Polymers

Commercial polymers (plastics, fibres, etc.) frequently contain monomeric substances (additives and impurities): plasticisers, antioxidants, UV-stabilisers, dyes, optical brighteners, lubricants, antistatics (additives), residual monomer, traces of catalyst, degradation products and dissolved atmospheric pollutants (impurities).

Many of these compounds absorb strongly nUV/VIS-radiation and thus disturb the investigation of the polymer, even if present only in traces. For the same reason, UV-absorption is a suitable method for analysing these substances, preferentially after extraction with a solvent which does not dissolve the polymer or by repeated solution/precipitation yielding pure solid polymer and additives dissolved in the solvent/non-solvent mixture.

Some additives have been designed to give optimum absorption in the nUV, e.g. the 2-hydroxybenzophenones and 2-(2-hydroxyphenyl) benzotriazoles used as screening type UV-stabilisers[18]. Polymer spectroscopy in its strict sence requires pure polymer samples so that carefully prepared and purified laboratory products will in general be preferred. Several polymers, however, (e.g. PP) can be prepared in the laboratory only with difficulties if special requirements concerning tacticities, molar mass, etc. have to be met. In such cases, purification of commercial polymers is a feasible way of obtaining suitable samples. Impurities which are chemically linked to the polymer cannot be removed by the usual solvent techniques, but in special cases chemical purification (e.g. hydrogenation of $C=C$ and $C=O$ groups) may be successful.

If the polymer contains electron-donating (D) or -accepting (A) groups, monomeric compounds with the opposite property will give rise to formation of DA complexes[29]. The corresponding charge transfer absorption bands are due to a partial shift of electrons from the donor to the acceptor, in most cases according to Eq. (3.12).

$$\{-D \cdot A \xrightarrow{\ h\nu\,CT\ } \{-(D^+ \cdot A^-)^*. \tag{3.12}$$

The resulting bands are often in the VIS and thus cause a specific colouring of the polymer solution or film. It should be noted that the optical density of the CT band in DA solutions does not obey the simple Lambert Beer's law, since formation and dissociation of the complexes depend on the concentrations and, more exactly, on the activities of D and A and on the equilibrium constant of complex formation[30].

The ionisation energy and electron affinity in polymers have already been discussed in Sect. 2.5.2, see also Fig. 2.10. Strong electron donors and acceptors, either polymers or monomers, combine to give radical-ion complexes in which the charge transfer occurs in the electronic ground state [Eq. (3.13)]. The radical ions formed and/or their

Table 3.3. Information obtained by UV/VIS absorption relating to structure and dynamics of polymeric systems

Structure		Dynamics	
Chemical structure	+	Movements of the chain, seg-	−
Tacticity	−	ments and sidegroups	
Conformation	+	Phonons	−
Crystallinity	(+)	Excitons	+
Electronic structure	+	Complex formation and related	+
		phenomena	

degradation products often show characteristic absorption bands in the VIS range, sometimes extending to the nIR.

Both phenomena – charge transfer in the ground and the excited state – play an important role in the electrical conductivity of polymers[31]. If the absorption coefficients of the DA complex or of the radical ions

$$\{-D + A \;\rightarrow\; \{-D^{\cdot\,+} + A^{\cdot\,-} \,.$$

(3.13)

are known, nUV/VIS/nIR-absorption spectroscopy can be used in order to analyse the amount and nature of the polymeric complex(es) formed.

3.3 Résumé of nUV/VIS-Absorption Spectroscopy

The following discussion includes certain aspects of the fUV (π electron absorption) and the nIR (electronic bands extending into this range. Table 3.3 shows the essential fields of application. Analytical applications are the identification of aromatic polymers, copolymers and additives. A strong conformational effect has been shown to occur in DNA hypochromy, and weaker effects of this kind may be used to explore the conformation of aromatic synthetic polymers in solution. Polymer crystallinity is another field not yet studied by UV/VIS-absorption, although specific splitting at low temperature is in principle to be expected in crystalline polymers. The electronic structure is studied by UV/VIS [and by ESCA (Chap. 2) and luminescence spectroscopy (Chap. 4)]. If the absorbing groups are closely spaced, the excited states may migrate as excitons, as will be shown in Chap. 4. Finally, the formation of DA complexes is an important application of nUV and especially VIS (+nIR) absorption spectroscopy.

The nUV/VIS-absorption spectroscopy seems to be a very important tool for studying
– aromatic and heteroaromatic polymers
– photochemistry of polymers
– degradation of polymers
– polymeric DA complexes and radical ion complexes.

References

1. Partrigde, R.H.: J. Chem. Phys. *45*, 1685 (1961)
2. Buck, W.C., Thomas, B.R., Weinreb, A.: J. Chem. Phys. *48*, 549 (1968)

3. Partrigde, R.H.: J. Chem. Phys. *47*, 4223 (1967)
4. George, R.A., Martin, D.H., Wilson, E.G.: J. Phys. C: Solid State Phys. *5*, 871 (1972)
5. George, R.A., Roberts, I., Wilson, E.G.: J. Phys. E: Sci. Instrum. *4*, 384 (1971)
6. Bloor, D.: Chem. Phys. Lett. *40*, 323 (1976)
7. Koch, E.E., Sonntag, B.F.: Molecular Spectroscopy with Synchrotron Radiation in Synchrotron Radiation, C. Kunz (ed.), Topics in Current Physics, Berlin, Heidelberg, New York: 1979, p. 269
8. Jortner, J, Leach, S. (eds.): Perspectives of Synchrotron Radiation. Application to Molecular Dynamics and Photochemistry. J. Chim. Phys 77, No. 1 (1980) pp. 1–57
9. Lee, C.H., Waddell, W.H., Casassa, E.F.: Macromolecules *14*, 1021 (1981)
10. Kortüm, G.: Kolorimetrie, Photometrie und Spektrometrie, 4th Ed., Berlin, Göttingen, Heidelberg: Springer 1962
11. Murrell, J.N.: The Theory of Electronic Spectra of Organic Molecules, London: Methuen 1963; German Transl.: B.I. Hochschultaschenbücher, No. 250/250 a, Mannheim: Bibliogr. Inst. 1967
12. Herzberg, G.: Molecular Spectra and Molecular Structure I. Spectra of Diatomic Molecules, 2nd Ed., New York: Van Nostrand-Reinhold 1950
13. Kasha, M.: Discuss. Faraday Soc. *9*, 14 (1950)
14. Buck, W.L., Thomas, B.R., Weinreb, A.: J. Chem. Phys. *48*, 549 (1968)
15. Schauenstein, E., Klöpffer, W.: Acta histochim. Suppl. VI, 227 (1965)
16. Klöpffer, W.: Europ. Pol. J. *11*, 203 (1975)
17. Kubica-Kowal, J.: Makromol. Chem. *178*, 3017 (1977)
18a. Geuskens, G. (ed.): Degradation and Stabilisation of Polymers, London: Applied Science Publ 1975
18b. Rånby, B, Rabek, J.F.: Photodegradation, Photooxidation and Photostabilisation of Polymers, London: Wiley 1975
18c. Hawkins, W.L. (ed.): Polymer Stabilization, New York: Wiley 1972
19. Klöpffer, W.: J. Chem. Phys. *50*, 2337 (1969)
20. Magazanik, B., Chargaff, E.: Biochem. Biophys. Acta 7, 396 (1951)
21. Thomas, R.: Biochem. Biophys. Acta *14*, 231 (1954)
22. Tinoco, I., Jr.: J. Am. Chem. Soc. *82*, 4785 (1960)
23. Eisinger, J.: Photochem. Photobiol. 7, 597 (1968)
24. Bolton, H.C., Weiss, J.J.: Nature *195*, 666 (1962)
25. Nesbet, R.K.: Mol. Phys. 7, 211 (1964)
26. Fowler, G.N.: Mol. Phys. *8*, 383 (1964)
27. Vala, M.T., Rice, S.A.: J. Chem. Phys. *39*, 2348 (1963)
28. Brüssau, R.J., Stein, D.J.: Ang. Makromol. Chem. *12*, 59 (1970)
29. Pearson, J.M., Turner, S.R., Ledwith, A., in: Molecular Association, R. Foster (ed.), London: Academic Press 1979, p. 79
30. Briegleb, G.: Elektronen-Donator-Acceptor-Komplexe, Berlin, Göttingen, Heidelberg: Springer 1961
31. Mort, J., Pfister, G. (eds.): Electronic Properties of Polymers, New York: Wiley 1982
32. Hashimoto, S., Seki, K., Sato, N., Inokuchi, H.: J. Chem. Phys. *76*, 163 (1982)
33. Armstrong, D.R., Jamieson, J., Perkins, P.G.: Theor. Chim. Acta *50*, 193 (1978)
34. Schauenstein, E., Bayzer, H.: J. Pol. Sci. *16*, 45 (1955)
35. Beavan, S.W., Hargreaves, J.S., Phillips, D.: Advances in Photochemistry Vol. *11*, 207, New York: Wiley 1979
36. Rao, C.N.R.: Ultraviolet and Visible Spectroscopy, London: Butterworth 1961
37. DMS UV Atlas of Organic Compounds, London: Butterworth, Weinheim: Verlag Chemie 1966–1971
38. Friedel, R.A., Orchin, M.: Ultraviolet Spectra of Aromatic Compounds, New York: Wiley, London: Chapman & Hall 1951
39. Pestemer, M., Scheibe, G., Schöntag, A., Brück, D., in: Landolt-Börnstein, 6. Auflage, I. Band, 3. Teil (Molekeln II), 78, Berlin, Göttingen, Heidelberg: Springer 1951
40. Pestemer, M.: Correlation tables for the structural determination of compounds by ultraviolet light absorptiometry, Weinheim: Verlag Chemie: 1974
41. Calvert, J.G., Pitts, J.N.P., Jr.: Photochemistry, New York: Wiley 1966

35

42. Phillips, J.P., et al. (eds.): Organic Electronic Spectra Data, Vol. I–VIII, New York: Wiley 1973
43. Sadtler Standard Ultraviolet Reference Spectra, London: Heyden
44. Bloor, D., Williams, R.L., Ando, D.J.: Chem. Phys. Lett. *78*, 67 (1981)
45. Sixl, H.: Spectroscopy of the Intermediate States of the Solid State Polymerization Reaction in Diacetylene Crystals, in Advances in Polym. Sci. (1983)
46. Takai, Y., Ozawa T., Mizutani, M. Ieda, M.: J. Pol. Sci. Pol. Phys. Ed. *15*, 945 (1977)
47. Davydov, A.S.: Theory of Molecular Excitons, translated by M. Kasha and M. Oppenheimer Jr., New York: McGraw-Hill, 1st Ed. 1962, 2nd Ed. 1971
48. García-Rubio, L.H.: J. Appl. Pol. Sci. *27*, 2043 (1982)
49. Ouchi, I.: Polymer (Japan) *15*, 225 (1983)

4 Fluorescence – and Phosphorescence Spectroscopy of Polymers

4.1 Radiative and Radiationless Transitions

4.1.1 The Jablonski Diagram

This section is complementary to the foregoing one, since the luminescence of organic molecules is due to radiative transitions between the same electronic levels, but in the reverse sense, compared to absorption. Furthermore, (delayed) emission can be observed by forbidden transitions which are not observable in absorption.

Luminescence is not observed, however, if there are efficient radiationless processes competing with the emission of photons.

The important energy levels of poly-atomic molecules – in our case of the emitting polymer chromophores – are usually presented in the form of a Jablonski-Diagram, choosing the singlet ground state S_0 as the common baseline (Fig. 4.1). The transitions $S_0 \rightarrow S_1, S_2 \ldots S_n$ symbolise the UV/VIS-absorption processes discussed in Chap. 3 (see also the discussion of energy level diagrams, Sect. 3.2.2.2).

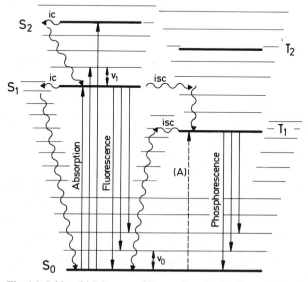

Fig. 4.1. Jablonski-Diagram of Energy Levels of Polyatomic Molecules in Condensed Phase

4.1.2 Fluorescence

Fluorescence occurs with only few exceptions from S_1, even if in absorption primarely a higher excited state S_n has been created (Fig. 4.1). This behaviour is also observed in the case of phosphorescence ($T_1 \rightarrow S_0$) and is known as Kasha's rule [1]. The reason for the absence (or extreme faintness) of emissions from higher excited states to lower excited ones or to the electronic ground state consists in the rapid internal conversion (Sect. 4.1.3) competing with luminescence [2]. In the singlet system, S_1 in its vibrational ground state offers a metastable intermediate in the deactivation cascade starting at S_n or at vibrationally excited S_1. The fluorescent transition $S_1 \rightarrow S_0$ is a dipole transition which is subject to the selection rules discussed in Chapt. 3. With the exception of lasing systems, not yet described for polymers, fluorescence occurs spontaneously, the rate of emission ($k_e = 1/\tau_e$) being a measure of the allowedness of the transition and thus proportional to $f_{0,1}$ ($f_{0,1} = 1$ roughly corresponds to $k_e = 10^9 s^{-1}$).

The vibrational structure of fluorescence depends on the vibrations (v_0) in S_0 coupling with the emission and is often similar to that of long wavelength absorption ($v_1 \approx v_0$, "mirror image" of fluorescence and absorption).

4.1.3 Internal Conversion and Inter-System Crossing

Internal conversion (ic) denotes a radiationless transition between two electronic states of the same multiplicity, intersystem crossing (isc) a similar radiationless transition connected with a change in multiplicity (mostly $S_1 \rightarrow T_n$ or $T_1 \rightarrow S_0$, see Fig. 4.1). Radiationless transitions are isoenergetic transitions to a vibrationally excited lower electronic state. In condensed phases (the only ones that are of interest to the spectroscopy of polymers) these transitions are followed by very rapid vibrational relaxation, i.e. transfer of vibrational energy to the surrounding medium. The rate of radiationless transitions decreases (roughly) exponentially with increasing energy difference between the electronic states involved [3] and thus explains Kasha's rule. The rate is furthermore decreased by the spin selection rule, which also applies to radiationless transitions so that isc, all other conditions being equal, is slower than ic.

4.1.4 Phosphorescence

The metastable state responsible for phosphorescence (T_1) is long-lived since the highly forbidden spin reversal is needed for both the radiative (phosphorescence) and the ratiationless (isc) deactivation to S_0 [4]. Whereas fluorescence decays with $\tau \approx 1$ to 100 ns, phosphorescence is much slower: $\tau \approx 1$ ms to 10 s (see also Sect. 4.3). Furthermore, phosphorescence is always observed at longer wavelengths compared to fluorescence; this is due to energy lowering by electron exchange $E(T_1) < E(S_1)$. The vibrational quanta v_0 coupling with $T_1 \rightarrow S_0$ may be different from those observed in the fluorescence spectrum. Due to the extremely long lifetime of T_1, this excited state is highly susceptible to quenching processes (e.g. by O_2) so that phosphorescence can in general only be observed at low temperature and in rigid matrices when diffusion is suppressed.

4.2 Experimental

Fluorescence and phosphorescence spectra are recorded by spectrofluorimeters [5, 40], as shown in Fig. 4.2. The radiation source emitting a continuum in the nUV/VIS-range

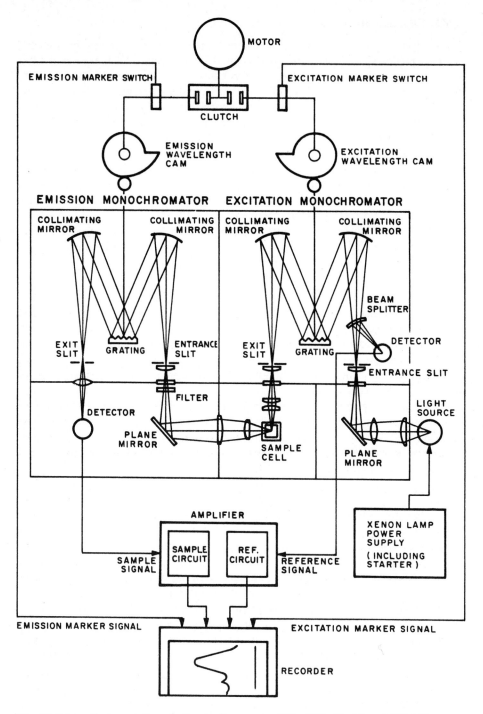

Fig. 4.2. Schematic presentation of a Spectrofluorometer (Hitachi/Perkin-Elmer MPF A4)

Table 4.1. Fluorescence quantum efficiencies of reference compounds[a]

Substance	T (K)	η_F	Ref.
Quinine sulphate (1.0 N H_2SO_4)	298	0.55	[8]
9,10-Diphenyl anthracene (cyclohexane)	298	1.00	[8, 9]
9,10-Diphenyl anthracene (ethanol)	298	0.94	[8]
	77	1.00	
Anthracene (crystalline)	298	0.9	[11]

[a] Performance of relative measurements of quantum efficiencies see Parker[6]

is a stabilised Xenon arc. The first monochromator is used to select a fixed wavelength for recording emission spectra. If luminescence excitation spectra are recorded, the excitation wavelength is varied at fixed emission wavelength. The first monochromator can be replaced by a suitable filter if only emission spectra are to be recorded.

The sample is contained in quartz tubes or cuvettes if in the form of solution, powders or granular materials.

Polymer films produced by slow evaporation of solutions onto a quartz support are mounted on a solid sample holder.

The solvent used to produce the films has to be of high purity. Low-temperature fluorescence and phosphorescence spectra are recorded using samples in quartz tubes or on a rod immersed in liquid nitrogen in a quartz Dewar vessel. Measurements performed below 77 K or requiring temperature variation require special cryostats.

The phosphorescence is recorded using rotating mechanical devices (sector discs or cans) separating the fast decaying fluorescence from the slow phosphorescence.

The second monochromator spectrally resolves the luminescence originating from the sample and can be replaced by suitable filters for recording only excitation spectra.

The detector nearly always is a photomultiplier whose sensitivity maximum should coincide with the emission of the sample in measurements requiring high sensitivity. In modern spectrometers the output of the detector is corrected for equal photon intensity in order to obtain "true" emission spectra; otherwise, correction curves can be obtained by standard lamps or using reference compounds of known spectral properties.

The most important sources of errors are:

- Scattered light from the source, especially in fluorescence measurements (the phosphoroscope acts as a trap of scattered light), if light sources with strong spectral lines and if strongly scattering samples are used, e.g. crystalline polymers.
- Smekal Raman effect (see Sect. 6.2), can be detected and eliminated by changing the excitation frequency.
- Chemical sources of error: impurities in solvent and sample; reabsorption and quenching in concentrated solutions.

B. Electronic Spectroscopy

Quantum efficiencies (see Sect. 4.3) (Table 4.1) are determined relative to known ones of reference substances using identical conditions of excitation and detection for sample and reference. This apparently simple comparative method has several sources of error[6-10], especially in the case of films (reabsorption of fluorescence, different scattering in sample and reference, etc.).

In addition to spectral distribution and quantum efficiency of luminescences, their decay behaviour – in the simplest case an exponential decay according to Eq. (4.1) – is of paramount importance in order to interpret and understand polymer emissions.

$$I(t) = I(0) \; e^{-t/\tau}. \tag{4.1}$$

Decay times can be estimated from quenching experiments or, better, measured directly using short pulses of exciting radiation and following up the decay of the luminescence to be studied. The slowly decaying phosphorescence can be interrupted using a mechanical shutter ($\tau \geqslant 1$ ms) and recorded by means of an oscillograph or recorder ($\tau \geqslant 1$ s).

The fluorescence decay in the range $\tau \approx 1$ ns to 1 μs can be studied using flashlamps and a sampling oscillograph, much more reliably, however, by means of the single photon counting technique[12, 56] and multichannel analysers. This technique has been extended to give time resolved emission spectra[13] and recently has been used to study polymer solutions.

Ultrashort fluorescences have been measured by means of a laser pulse technique known as "picosecond spectroscopy"[14], the most advanced tool for measuring the decay of emissions in the ps to ns range being the streak camera[15].

4.3 Quantum efficiency, decay time and rate constants

In polymers, the emitting chromophore is often not identical with the absorbing one, as implied in the discussion of the Jablonski Diagram in Sect. 4.1. The absorbing group may be reversibly converted into another with different properties, or the electronic excitation energy may be transferred to other groups which eventually are radiatively deactivated. The quantum efficiency as experimentally observed by comparison with a standard (η) may therefore be quite different from the "true" quantum efficiency (ϕ) which is a characteristic feature of the emitting group, irrespective of the mode of excitation.

For the simplest case, we consider an absorbing—emitting group and neglect external quenching, the notation of the rate constants is explained in the simplified Jablonski Diagram (Fig. 4.3). Under these simple conditions, the true quantum efficiencies of flu-

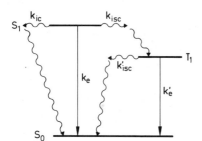

Fig. 4.3. Simplified Jablonski-Diagram. Vibrational relaxation (vertical wave lines) is assumed to be much faster than the other processes (condensed phase)

orescence (ϕ_F) and phosphorescence (ϕ_P) are given by the ratios of the radiative rate constants to the sum of all competing rate constants [Eqs. (4.2) and (4.3)] [14]

$$\phi_F = \frac{k_e}{k_e + k_{ic} + k_{isc}},$$ (4.2)

$$\phi_P = \frac{k_e'}{k_e' + k_{isc}'}.$$ (4.3)

The true quantum efficiency may generally be defined according to Eq. (4.4) and

$$\phi = \frac{k_{rad}}{k_{rad} + \Sigma k_{nrad}}$$ (4.4)

indicates the fraction of emitting states radiatively deactivated. The experimental quantum efficiencies are related to the number of photons absorbed and are identical with the true quantum efficiencies if each photon absorbed creates one emitting state. For example, $\phi_F = \eta_F$ if the absorption creates S_1, or, if higher excited singlets S_n are created in a first step and there are no side reactions, as ic to S_0, isc to T_n and photochemical reactions from S_n states. This condition is often fulfilled, but cannot be expected from the start. The true quantum efficiency of phosphorescence is identical with the experimental one (using, as usual, excitation via the singlet system) only if $k_{isc} \gg k_e + k_{ic}$, i.e. $\phi_{isc} \approx 1$. This behaviour is often observed in carbonyl containing chromophores. In general Eq. (4.5) applies, provided there are no

$$\eta_P = \phi_P \times \phi_{isc}$$ (4.5)

side reactions before T_1 reached.

As can be seen from Eqs. (4.2)–(4.4), the quantum efficiencies give only ratios of rate constants. In order to determine the rate constants, we need additionally the decay times. In Eq. (4.1) (exponential decay), the decay time τ is the reciprocal of the sum of all rate constants determining the decay of the excited states S or T [Eqs. (4.6) and (4.7)].

$$\frac{1}{\tau_S} = k_e + k_{ic} + k_{isc},$$ (4.6)

$$\frac{1}{\tau_T} = k_e' + k_{isc}'.$$ (4.7)

In calculating rate constants from measured τ and η data, it is often assumed that $k_{ic} \ll k_e + k_{isc}$ so that $\phi_{isc} \approx 1 - \phi_F$. This assumption facilitates the evaluation but several cases are known (e.g. the constituent bases of DNA and biphenylene) where ic is the fastest deactivation process of S_1.

4.4 Fluorescence in Polymers [15]

4.4.1 Isolated and Crowded Fluorescent Groups

In studying the fluorescence of polymers it is useful to distinguish between isolated fluorescent groups, e.g. in statistical copolymers and fluorescent groups in homopolymers or blockcopolymers. In the first case, only weak interactions between the excited group and its surrounding are in general to be expected, whereas in the second case strong interactions may occur, especially if the groups are favourable spaced and oriented relative to each other.

Molecules in excited states have drastically different properties compared to the same molecules in their electronic ground state [18]: their ionisation energy is lower ($I_g^* = I_g - E_{S_1}$ or $-E_{T_1}$), their electron affinity is higher ($A_g^* = A_g + E_{S_1}$ or $+E_{T_1}$), their acid/base properties are different due to different electron distribution, the dipole moment is different and so is the general chemical reactivity (photochemistry!). It can therefore easily be understood that interactions of excited groups play a greater role than in the ground state. The most conspicuous interaction observed in polymers leads to the formation of excimers [16, 17], complexes existing in the excited state only; these can be recognised by a strong red shift of fluorescence, compared to monomer emission. Energy transfer by exciton hopping – a random electronic diffusion of S_1 or T_1 – is another important process in aromatic polymers to be studied by luminescence spectroscopy. The joint occurrence of exciton hopping and excimer formation is the typical combination of processes competing with simple deactivation in these polymers [17]. This is especially true for solid aromatic polymers and liquid solutions. Rigid dilute solutions, on the other hand, behave apparently more like monomers, although weak interactions between the chromophores can be identified even in this case [19].

4.4.2 Fluorescence from Isolated Chromophores

Statistical copolymers containing a lower energy (S_1) component (*underlined*) in small concentration can be prepared from couples of suitable monomers, *styrene*/methyl methacrylate, *vinylnaphthalene*/styrene, *vinyl carbazole*/styrene, etc. The chemical bonding in the macromolecules induces some broadening of the fluorescence spectra which otherwise are very similar to those of suitable monomeric models. The fluorescent chromophor can be excited either directly, since the host polymer (majority component of the copolymer) is transparent in the region of $S_0 \rightarrow S_1$ absorption of the minority component, or by energy transfer via the majority component (Fig. 4.4). In the latter case, the mole ratio should exceed, say, 1/100.

Similar cases are trace impurities, e.g. residual chemical monomer and degradation products formed during extrusion, storage, use or weathering of polymers. Chemically linked fluorescent groups can easily be distinguished from dissolved ones by repeated dissolution/precipitation or by GPC.

Isolated fluorescent groups represent the monomer aspect of polymer spectroscopy and may be used for analytical purposes.

Collections of reference sprectra are given in the bibliography [7, 44, 45].

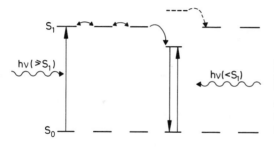

$hv(>S_1)$

$hv(<S_1)$

Fig. 4.4. Fluorescence Excitation of the minority component in a copolymer. The dashed arrow indicates rapid quenching if S_1 of the minority component is higher than S_1 of the majority component

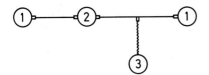

Fig. 4.5. Schematic view of different positions of fluorescence labels in a polymer chain

Fluorescence labels constitute a special case of isolated fluorescent groups in polymers [20-22, 41-43]. These groups, e.g. reactive dyes or aromatic molecules, are introduced into polymers in order to study the dynamics or orientation of the macromolecules.

Depending on the type of movement to be studied, the label may be fixed in mainly three different positions of the main chain (Fig. 4.5): at the end groups, firmly locked into the backbone or flexibly connected with it.

In order to investigate the dynamics of polymer chains, the labels are excited using polarised light, the fluorescence emitted by the label being observed through a second polariser. In a sample containing randomly distributed labels, the polarised light excites selectively those labels whose transition dipole moment is parallel to the electric vector of the excitation beam (photo selection). If the emitting state is identical with the absorbing one – or at least has the same transition vector – the degree of polarisation can only change if the label moves during τ. The time scale of this experiment is thus determined by the fluorescence decay time of the label. In more sophisticated experiments the depolarisation is not measured in a stationary manner but rather in a time-resolved way [15, 22]. It is a prerequisite of all these studies that competing depolarising processes, as energy transfer, are absent; consequently, the concentration of the labels has to be low (see Sect. 4.4.3).

In the absence of molecular movements, the polarisation of fluorescence can be used for studying the degree of orientation in solid polymers, e.g. drawn fibers and films. In this case [20], physically dissolved fluorescent molecules can be used as probes, introduced into the sample before stretching by diffusion or mixing before extrusion.

4.4.3 Energy Transfer between Isolated Groups or Dissolved Molecules

If the fluorescence spectrum of a polymer molecule or isolated chromophore of a polymer (D = energy donor) [1] overlaps with the absorption spectrum of another molecule or chromophore (A = energy acceptor), energy transfer may occur from excited D* to A, Eq. (4.8). This energy transfer, which does *not* involve emission and re-absorption of photons, is a non-radiative process first observed by Perrin [23] and theoretically explained by Förster [14, 24, 6] using dipole resonance as the main mechanism.

$$D^*(S_1) + A(S_0) \rightarrow D(S_0) + A^*(S_n). \tag{4.8}$$

The rate constant of energy transfer decays with the 6th power of the distance R between D and A.

---

1 Note that D and A have a different meaning than in Sect. 3.2.6 where electron donor-acceptor (CT) complexes have been discussed

By definition, the "critical radius", R_0 is the distance between D and A at which the probability of energy transfer es equal to the probability of all other deactivating processes of D*.

According to Förster's theory, R_0 can be calculated from spectroscopic data, Eq. (4.9)

$$R_0^6 = \frac{9,000 \ln 10 \varkappa^2}{128 \pi^5 N_L n^4} \phi_F(D)\Omega \tag{4.9}$$

$\varkappa^2$ (orientational factor) ≈ 0.6 in solids

N_L Loschmidt's number

n Refractive index of the polymer or solvent in the spectral region of fluorescence (D)/absorption (A) overlap

Ω Overlap integral according to Eq. (4.10)

$$\Omega = \int_0^\infty f_D(v')\varepsilon_A(v') \frac{dv'}{v'^4}$$
$$\approx \sum_{v'_1}^{v'_2} f_D(v')\varepsilon_A(v') \frac{\Delta v'}{v'^4}, \tag{4.10}$$

where the donor's fluorescence spectrum f_D is normalised such that

$$\int_0^\infty f_D(v')dv' = 1.$$

The integral in Eq. (4.10) can be approximated by a summation over the range of measurable overlap (v'_1 to v'_2) using appropriate steps $\Delta v'$ which depend on the vibrational structure displayed by D and A.

The measurement of R_0 for suitable D/A pairs offers a spectroscopic method of measuring distances in the range of 0.5 up to 10 nm in the case of extremely strong overlap, high ϕ_F and small $\bar{v}'$. Using suitably D and A labelled end groups, this method could be used for measuring the end-to-end distance in polymer coils.

Calculations of energy transfer are frequently necessary in order to make comparisons with experimental measurements or as an alternative to the experiment if, e.g., selective excitation of D is not possible. Transparent polymers have frequently been used as rigid solvents for energy transfer studies of dissolved monomeric compounds.

4.4.4 Singlet Excitons

In homopolymers or copolymers with a high content of fluorescent groups the first excited singlet state is not fixed to the group where it has originally been created by absorption of photon or some other excitation process (X-rays etc.) but rather moves [26] around in a random hopping process until deactivation occurs by emission of fluorescence, radiationless deactivation or trapping and quenching. The reason for this singlet exciton hopping process is a step-by-step Förster mechanism (4.4.3) although at the close distance between nearest neighbours higher multipole and electron exchange terms beside dipole-dipole may contribute to the rate of the elementary transfer step [25]. Essentially, the transition dipole moment (**M**) induces the same moment at a neighbouring group so that the energy is transferred without emission and absorption of a

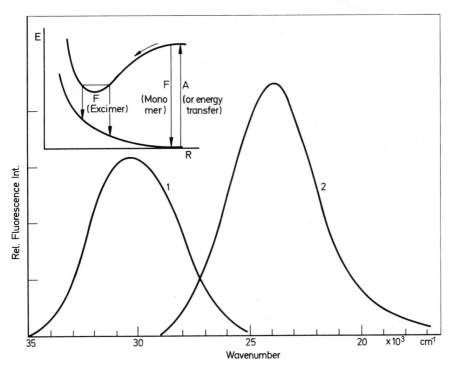

Fig. 4.6. Excimer fluorescence of PS (1) and PVCA (2) to be compared with absorption spectra in Fig. 3.3. Insert: simplified excimer potential scheme

photon within a time of about 1 ps, depending on $|\mathbf{M}|$, distance and relative orientation of the groups. It should be remembered that wavelike (coherent) excitons have not yet been detected in polymers (Sect. 3.2.5).

In the absence of traps monomer – like fluorescence would be expected in aromatic homopolymers. This, however, has been observed only in rigid solutions, although a small red shift is detected in most cases [19]. In liquid solutions and in solid films of aromatic polymers the monomer fluorescence is in general replaced by a broad, strongly red-shifted and structureless excimer fluorescence [15–17, 26], Fig. 4.6. For sterical reasons, excimer formation can take place only at a few sites of higher than average energy, since these complexes, which can exist only in the excited state [27], require nearly complete overlap of the molecular planes in a sandwich-like arrangement. Excimer fluorescence and the absence of monomer fluorescence in pure solid aromatic polymers proves energy transfer since absorption of photons occurs randomly, whereas fluorescence is emitted, only from specific centres, called excimer-forming sites [46]. Since furthermore excimers do not have a bonding ground state, they cannot act as acceptors in a long-range Förster-transfer so that energy transfer must be due to excitons.

In solid films energy transfer can quantitatively be studied using quenching or sensitised fluorescence of guest substances dissolved in the polymer to be studied. Equations (4.11) and (4.12) are modified Stern-Volmer equations [26, 46],

$$Q = \frac{I_0 - I}{I} = \frac{C_G}{C_E} \tag{4.11}$$

Q Quenching factor

I_0, I Fluorescence intensity without and with guest molecule of

C_G mol guest/mol basic unit of polymer

C_E mol excimer-forming sites/mol basic unit of polymer.

Equation (4.11) is valid if monomer fluorescence is completely quenched by energy transfer to excimer-forming sites: in typical aromatic polymers, $c_E \approx 10^{-3}$ to 10^{-2} mol/mol basic unit [26]. The reciprocal of C_E gives the average number of exciton jumps in the polymer which is necessarily lower than in molecular crystals where Eq. (4.12) applies.

$$Q = n\, C_G \tag{4.12}$$

$n =$ number of exciton jumps $\approx 10^4$ to 10^5 in typical molecular crystals [26].

In a polymer free of excimer-forming sites Eq. (4.12) would be suitable for describing energy transfer by excitons.

The excimer lifetime is independent of C_G as long the excimer is not an intermediate step in energy transfer, but rather a passive trap. However, if a guest shows strong absorption in the region of excimer fluorescence and is present in high concentration, the excimer is bound to act as energy-donor in Förster transfer. This should be indicated in a steep decrease of the excimer decay curve near $t = 0$ [27]. Thus, kinetic measurements are needed in order to elucidate the details of energy transfer in aromatic polymer films. This is even more important in fluid solutions where the kinetics are complicated by the superposition of exciton and molecular dynamics, since in flexible chain molecules excimers can be formed during the lifetime of the excited state. Recently, evidence has beeen produced which contradicts simple Birks kinetics [2, 17] in polymer solutions [28].

Identifying the excimer fluorescence of a specific polymer chromophore often involves dimeric or trimeric model compounds which in studies aimed at clarifying the influence of tacticity on energy transfer, chain conformation, etc. should reflect the different configurations possible [16, 29]. Generally, isotactic configuration seems to favour excimer formation in polymers and oligomeric models [47].

4.4.5 Excimer Fluorescence as a Probe in Polymer Studies

Excimer fluorescence is a useful probe in polymer studies owing to the structural requirements of excimer formation, implying close contact of two groups, either frozen in, as in solids, or dynamically formed as possible only in liquids. Since in liquids energy transfer to preformed sites and dynamic excimer formation are possible, the kinetic interpretation of fluorescence experiments may be difficult. In order to use excimer fluorescence as a probe in liquids, the excimer-forming groups have to be present in dilute form, as recently reported by Morawetz [30], using a copolymer of PA [66]:

As a dimeric model, the N,N′ diacetyl derivative of the phenyltype excimer probe has been measured as a function of temperature. Excimer formation has been found to increase with increasing temperature, in accordance with expectations, if internal hindrance was responsible for low excimer yield at low temperature. The copolymer showing exactly the same temperature dependence of excimer relative to monomer fluorescence intensity (I_E/I_M) indicates the absence of "crankshaft-like" motions in this – and probably other – polymer chains, involving the concerted rotation of two bonds.

Important information on chain dynamics can be obtained from endgroup labeled macromolecules using excimer-forming pyrene groups as labels, as shown in the pioneering work by Carla Cuniberti[48] and subsequent time-resolved measurements of very narrow fractions of Pyr-Ps-Pyr, performed by Winnik and coworkers[49, 50]. Measurements of this type can be used in order to test advanced theories of intramolecular dynamics and end-group cyclization[51–53].

A second example in which excimer fluorescence is applied as a probe is the study of compatibility in polyblends of poly (2-vinyl naphthalene) in a series of solid polyalkylmethacrylates[31]. C. W. Frank prepared dilute solid solutions of P2VN in a series of polyalkylmethacrylates showing different alkyl substituents and correlated the calculated solubility parameters of host and guest polymer with measured fluorescence intensity ratios I_E/I_M of P2VN. This ratio was found to be minimal – indicating good compatibility – if the solubility parameters of P2VN and host polymer are identical. Pure excimer fluorescence would be expected for complete seggregation, i.e. formation of little particles of P2VN (Sect. 4.4.4) in the host matrix. The merit of this method, however, lies in the detection of the early stages of seggregation, not possible by other methods. Recently, this work has been extended, e.g. to the phase separation of PS/ Poly(vinylmethyl ether) blends[54].

4.5 Phosphorescence in Polymers

4.5.1 The Phosphorescent Triplet State

The phosphorescent state, T_1 (Sect. 4.1.4) is distinct from its corresponding singlet state (mostly S_1 or S_2) with respect to energy, lifetime and spin state. Due to its paramagnetism it can also be studied by ESR (Sect. 9.7) and, as a result of its slow decay, the absorption spectrum ($T_1 \rightarrow T_n$) can be measured using intense pulse excitation. Since direct absorption[4] ($S_0 \rightarrow T_1, \dots T_n$) is very weak, T_1 is generally populated by isc via the singlet system. Alternatively, it may be formed by energy transfer (Sects. 4.5.2 and 4.5.3) or chemical sensitisation[32, 33], Eq. (4.13). The excited triplet aceton transfers its

$$\text{(4.13)}$$

Tetramethyl-
dioxetane

$(S_1 \text{ or } T_1) (S_0)$

energy to polymer chromophores of lower (PVCA) or slightly higher (PS) triplet energy.

47

4.5.2 Isolated Phosphorescent Groups

A phosphorescent group of a polymer without appropriate neighbours with which it may interact behaves essentially like a physically dissolved molecule. The phosphorescence of the statistical copolymer PS/VCA, e.g. shows a O—O peak which is red-shifted by 70 cm^{-1} only with regard to N-isopropyl carbazole as monomeric model. The vibrational resolution, however, is poorer in the polymer spectrum. Other examples of isolated phosphorescent groups are oxidatively formed chromophores in commercial polymers. The phosphorescence spectrum of PS shown in Fig. 4.7 is undoubtly due to acetophenone type end groups –CO–C$_6$H$_5$, as indicated by the characteristic vibrational structures in the excitation – and the emission spectra. The latter is dominated by C=O valence vibrations (Fig. 4.8) which in severely degraded PS samples can also be detected in the mIR spectrum. It can be seen from Fig. 4.7 how small the S_1–T_1 gap is in this group from the close position of the O—O peaks of phosphorescence excitation ($S_0 \rightarrow S_1$) and emission ($T_1 \rightarrow S_0$); this is one reason for the efficient isc preventing fluorescence in this group. The analytical usefulness of the spectrum shown is due to the high quantum yield (ϕ_P) and, above all, to the clearly resolved vibrational structure.

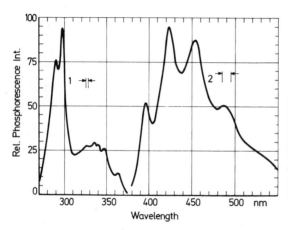

Fig. 4.7. Phosphorescence excitation (1) and emission (2) spectrum (not corrected) of a thick film of commercial polystyrene at 77 K. Excitation of (2) at 315–355 nm. The sample has not been artificially degraded. The structure in (1) near 290 nm is due to residual monomeric styrene

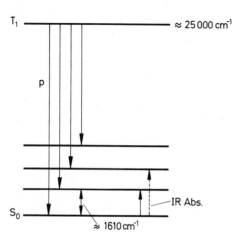

Fig. 4.8. Phosphorescence and IR absorption in acetophenone type end groups (see Fig. 4.7)

There are many polymer phosphorescences without vibrational structure which contain little analytical information. Careful comparison with possibly monomeric models and additional comparative measurements of the decay of polymer and monomer phosphorescence is clearly necessary in these cases if any conclusions on the chemical nature of the groups are to be drawn.

Energy transfer from isolated groups may occur over small distances only, since electron exchange[25] as the dominant transfer mechanism requires some overlap of electronic wave functions of D and A. Perrin's model[34] is used to describe triplet energy transfer[26, 35] quantitatively. Energy transfer to molecular (triplet) oxygen, the most common quenching mechanism, leads to electronically excited singlet oxygen which may react with double bonds of the macromolecules to yield hydroperoxides.

4.5.3 Triplet Excitons in Polymers

Evidence for mobile triplet states in polymers is derived from two sources; sensitised guest phosphorescence and T-T annihilation[19, 26, 32]. The smaller hopping rate compared to singlet excitons is compensated by the fact that the triplet exciton has much more time due to its slow decay. If the electronic excitation is strongly localised within the chromophore (small overlap) and the triplet lifetime is short, then the triplet excitons have only a short range or are even immobile. This seems to be the case in aliphatic polyketones[36].

T-T annihilation is elegantly detected by delayed fluorescence without the need of adding guest molecules.

DF is caused by the singlet channel of T-T annihilation:

$$T_1 + T_1 \rightarrow S_0 + S_n \rightarrow S_1 \rightarrow S_0 + h\nu\text{DF}. \tag{4.14}$$

The delayed fluorescence is in general spectrally identical with prompt fluorescence (for deviations in polymers see[26, 37, 38] but kinetically related to the slowly decaying triplets so that experimentally DF is observed together with the phosphorescence spectrum.

Delayed fluorescence, in addition to phosphorescence, has been observed in several aromatic polymers, both in glassy solution and in the form of solid films at low temperature[26]. Triplet excimers have been identified as exciton traps, characterised by redshifted structureless emission in analogy to singlet excimers, but apparantly of different conformation. Triplet exciton hopping seems to be much slower in amorphous polymers compared to molecular crystals, probably due to shallow T-exciton traps[55].

In dilute glassy solutions of aromatic homopolymers, triplet excimers have not been detected, but rather red-shifts of the order of several 100 cm^{-1}[19] indicating some trapping of energy even in this cace. Mobile triplet states are indicated by delayed fluorescence whose intensity strongly increases with the molar mass of the polymer or polymer fraction used. This effect is due to the confinement of the excitons in the pseudo-1D polymer chains from which the coils are formed. At low molar mass or irradiation intensity, statistically uneven distribution of the triplet excitons over the different macromolecules of the sample may occur, so that some coils at a given time have no exciton, others one, two, etc. Studies of this kind represent a good example of the polymer aspect and at the same time demonstrate the polymers as real, although not well ordered, 1D-systems.

4.6 Résumé of Fluorescence and Phosphorescence Spectroscopy

As can be seen from Table 4.2 the luminescence of polymers gives relevant information on several items.

The analytical application depends critically on elevated quantum efficiencies ($\phi \approx$ 0.1 or higher) and well developed vibrational structure. The method is very sensitive but difficult (scattering effects, re-absorption) if quantitative measurements have to be made especially in solid polymers.

Tacticity has an indirect influence on the emission spectra of polymers since it faours certain conformations leading to excimer formation. The crystallinity has so far not been studied in detail. The relationship of luminescences with electronic structure in ground and excited states (S_1 and T_1) is evident.

Molecular movements can be studied by fluorescence depolarisation techniques often requiring labelled polymers. Alternatively, the dynamics of excimer formation can be used to study chain movement [39, 48, 49]. Phonons should be observable in low-temperature luminescence spectra of highly crystalline polymers but little is known about this at present.

Luminescence methods are indispensable for the study of excitons. Complex formation, at least in the excited state (excimers and exciplexes) is also studied by fluorescence and phosphorescence methods.

Despite this wealth of possible applications, it should be noted that luminescence spectroscopy is no routine method and probably never will be. This is mainly to be attributed to the following facts:

– In many pure polymers there are no suitable groups (luminophores)
– There are many disturbances such as reabsorption, luminescent impurities, quenching by oxygen and other impurities.

In other words: the high degree of sensitivity and specificity of this method, making it a unique tool of polymer research, seems to prevent its broader use as a routine methods for analytical purposes.

Fluorescence and phosphorescence spectroscopy is especially useful in studies concerning

– aromatic, heteroaromatic and carbonyl-containing polymers
– photochemistry of polymers

Table 4.2. Information obtained by fluorescence and phosphorescence on structure and dynamics of polymeric systems

Structure		Dynamics	
Chemical structure	+	Movements of the chain,	+
Tacticity	+	segments and side groups	
Conformation	+	Phonons	–
Crystallinity	–	Excitons	+
Electronic structure	+	Complex formation and related phenomena	+

- photophysics of polymers, including photoconductivity and scintillation counting
- dynamics of macromolecules in solution
- compatibility and phase separation of polymer blends.

References

1. Kasha, M.: Disc. Faraday Soc. *9*, 14 (1950)
2. Birks, J.B.: Photophysics of Aromatic Molecules, London: Wiley 1970
3. Henry, B.R., Siebrand, W., in: Organic Molecular Photophysics, Vol. 1, J.B. Birks (ed.), London: Wiley 1973 p. 153 and Vol. 2, p. 303 (1975)
4. McGlynn, S.P., Azumi, T., Kinoshita, M.: Molecular Spectroscopy of the Triplet State, Englewood Cliff: Prentice Hall 1969
5. Shulman, S.S.: Fluorescence and Phosphorescence Spectroscopy: Physico-chemical Principles and Practice, Oxford: Pergamon Press 1977
6. Parker, C.A.: Photoluminescence of Solutions, Amsterdam: Elsevier 1968
7. Berlman, I.B.: Fluorescence Spectra of Aromatic Molecules, 2nd Edition, New York: Academic Press 1971
8. Heinrich, G., Schoof, S., Güsten, H.: J. Photochem. *3*, 315 (1974/75)
9. Ware, W.R., Rothman, W.: Chem. Phys. Lett. *39*, 449 (1976)
10. Huber, J.R., Mahaney, M.A., Mantulin, W.W.: J. Photochem. *2*, 67 (1973/74)
11. Wright, G.T.: Proc. Phys. Soc. (London) B *68*, 241 (1955)
12. Ghiggino, K.P., Robert, A.J., Phillips, D.: Time-Resolved Fluorescence Techniques in Polymer and Biopolymer Studies in Advances in Pol. Sci. Vol. 40 (Luminescence) Berlin, Heidelberg, New York: Springer 1981
13. Beavan, S.W., Hargreaves, J.S., Phillips, D., in: Adv. Photochem. Vol. *11*, 207, New York: Wiley 1979
14. Förster, Th.: Fluoreszenz organischer Verbindungen, Göttingen: Vandenhoeck und Ruprecht 1951; Reprint 1982
15. Morawetz, H., Steinberg, I.Z. (Eds.): Luminescence from Biological and Synthetic Macromolecules, Annals of the New York Academy of Sciences, Vol. 366 (1981)
16. Hirayama, F.: J. Chem. Phys. *42*, 3163 (1965)
17. Klöpffer, W.: Intramolecular Excimers in Organic Molecular Photophysics, J.B. Birks (ed.), London: Wiley 1973, Vol. 1, p. 357
18. Turro, N.J.: Modern Molecular Photochemistry, Menlo Park/California: Benjamin/Cummings Publ. 1978
19. Klöpffer, W.: Spectros. Letters *11*, 863 (1978)
20. Nishijima, Y., Teramoto, A., Hiratsuka, S.: J. Pol. Sci. A-2 *5*, 23 (1967)
21. North, A.M., Soutar, J.: J. Chem. Soc. Faraday Trans. I *68*, 1101 (1972)
22. Valeur, B., Monnerie, L.: J. Pol. Sci. Pol. Phys. Ed. *14*, 11 (1976)
23. Perrin, J.: Compt. rend. *177*, 469 (1923)
24. Förster, Th.: Disc. Faraday Soc. *27*, 7 (1959)
25. Dexter, D.L.: J. Chem. Phys. *21*, 836 (1953)
26. Klöpffer, W.: Energy Transfer, in: Electronic Properties of Polymers, J. Mort and G. Pfister (eds.), New York: Wiley 1982
27. Förster, Th.: Ang. Chem. *81*, 364 (1969)
28. Phillips, D., Roberts, A.J., Soutar, J.: J. Pol. Sci. Pol. Phys. Ed. *18*, 2401 (1980)
29. Bokobza, L., Jasse, B., Monnerie, L.: Eur. Pol. J. *13*, 921 (1977)
30. Morawetz, H.: Science *203*, 405 (1979)
31. Frank, C.W., Gashgari, M.A.: Macromolecules *12*, 163 (1979)
32. Turro, N.J., Kochevar, I.E., Noguchi, Y., Chow, M.-F.: J. Am. Chem. Soc. *100*, 3170 (1978)
33. Klöpffer, W., Turro, N.J., Chow, M.-F., Noguchi, Y.: Chem. Phys. Lett. *54*, 457 (1978)
34. Perrin, F.: Compt. rend. *178*, 1978 (1924)
35. Ermolaev, V.L.: Sov. Phys. Uspekhi *80*, 333 (1963) (Engl. Transl.)
36. David, C., Putman, N., Lempereur, M., Geuskens, G.: Eur. Pol. J. *8*, 409 (1972)
37. Rippen, G., Kaufmann, G., Klöpffer, W.: Chem. Phys. *52*, 165 (1980)
38. Kim, N., Webber, S.E.: Macromolecules *13*, 1233 (1980)

B. Electronic Spectroscopy

39. Cuniberti, C., Perico, A.: Eur. Pol. J. *16*, 887 (1980)
40. Zander, M.: Fluorometry, Berlin, Heidelberg, New York: Springer 1981
41. Chapoy, L.L., Du Pré, D.B.: Fluorescence Probe Methods, in: Methods in Experimental Physics, Vol. *16*, Polymers, Part A, R.A. Fava (ed.) New York: Academic Press 1980
42. Anufrieva, E.V., Gotlib, Y.Ya.: Investigation of Polymers in Solution by Polarized Luminescence, in: Adv. Pol. Sci., Vol. *40*, Berlin, Heidelberg, New York: Springer 1981
43. Anufrieva, E.V.: Pure & Appl. Chem. *54*, 533 (1982)
44. Schmillen, A., Legler, R.: Lumineszenz Organischer Substanzen, in: Landolt-Börnstein, Neue Serie Bd. *3*, Berlin, Heidelberg, New York: Springer 1967
45. Standard Fluorescence Spectra, Vol. 1–5, Sadtler Res. Laboratories, Philadelphia (1974/75)
46. Klöpffer, W.: J. Chem. Phys. *50*, 2337 (1969)
47. De Schryver, F.C., Moens, L., Van der Auweraer, N., Boens, N., Monnerie, L., Bokobza, L.: Macromolecules *15*, 64 (1982)
48. Cuniberti, C., Perico, A.: Eur. Pol. J. *13*, 369 (1977)
49. Winnik, M.A., Redpath, T., Richards, D.H.: Macromolecules *13*, 328 (1980)
50. Redpath, A.E.C., Winnik, M.A.: J. Am. Chem. Soc. *104*, 5604 (1982)
51. Wilemski, G., Fixman, M.: J. Chem. Phys. *60*, 866, 878 (1974)
52. Doi, M.: Chem. Phys. *9*, 455 (1975)
53. Perico, A., Cuniberti, C.: J. Pol. Sci. Pol. Phys. Ed. *15*, 1435 (1977)
54. Gelles, R., Frank, C.W.: Macromolecules *15*, 1486 (1982)
55. Klöpffer, W.: Chem. Phys. *57*, 75 (1981)
56. Demas, J.N.: Excited State Lifetime Measurements, New York: Academic Press 1983

Part C. Vibrational Spectroscopy

5 Vibrations of Polymers

5.1 Introduction

Vibrational transitions of macromolecules have been demonstrated to influence the UV/VIS-absorption and emission spectra where these transitions cause the vibrational structure of electronic bands (Chap. 4). In the case of fluorescence and phosphorescence spectra this vibrational structure corresponds to ground state vibrations coupling with the electronic transition. Only a small fraction of all possible vibrations, however, can be identified in electronic spectra due to restrictions in coupling efficiency and spectral resolution. The methods of vibrational spectroscopy to be discussed in the following chapters allow a much broader range of polymer vibrations to be studied and additionally do not depend on the presence of chromophores with the exception of Resonance Raman Spectroscopy.

Infrared absorption and Raman spectroscopies do not, however, indicate all vibrations theoretically possible in polymers [1] since each molecule has $3n-6$ fundamental vibrations, where $n =$ number of atoms forming the molecule. Thus, polymers should have a nearly infinite number of vibrational bands and no useful information could be obtained from the spectra. Actually, polymer spectra are only moderately more complicated than monomer spectra. This is especially true for amorphous polymers whose vibrational spectra often can hardly be distinguished from corresponding spectra of monomeric model compounds. The reasons are:

1. Coincidence of transition energies (degeneration), especially in symmetric molecules.
2. Inefficient coupling between distant parts of the molecule. This can be understood immediately in polymers, where often the vibrating groups are separated by thousands of bonds.
3. Selection rules to be met regarding the change in dipole moments (IR) or polarisabilities (Raman) during vibration.

5.2 The Harmonic Oscillator

Molecular vibrations are due to the elastic rather than rigid nature of chemical bonds which can be elongated or twisted; furthermore, the angles between the bonds can be changed periodically [2].

Since the potential energy of the non-vibrating molecule is at its minimum, any disturbance requires the uptake of energy. Using a mechanical analogue taken from the macroscopic world, e.g. spheres interconnected with springs, it can easily be demonstrated that the characteristic vibrational frequency (v_v) has to be high if the strength of the spring (chemical bond) is high and/or if the mass of the spheres (atoms, parts of molecules) is small. The classical mechanical relationship between these magnitudes is given by Eq. (5.1) [2] for the harmonic oscillator.

$$v_v = \frac{1}{2}\pi \sqrt{\frac{k}{\mu}} \tag{5.1}$$

where k = force constant and μ = reduced mass, in the case of two vibrating masses m_1, m_2: $\mu = m_1 \cdot m_2/(m_1 + m_2)$. Equation (5.1) implies that the restoring force is proportional to elongation and consequently that the potential energy is a function of elongation. In the case of a stretching vibration the function $E_{pot}(x)$ is parabolic, Eq. (5.2).

$$E_{pot} = E_{pot}(0) + 1/2kx^2 . \tag{5.2}$$

This potential can be valid only near the equilibrium position, since the binding force diminishes during elongation (x). In the classical picture which can easily be visualised for a two-particle system (Fig. 3.2), the molecule rests in equilibrium with a minimum energy $E_{pot}(0)$. Disturbing the system would induce a vibration whose amplitude would be proportional to the strength of the disturbance.

This simple picture contradicts the laws of quantum mechanics in two points: first, a rest position at $x = 0$ is impossible, since in this case the position and the energy of the system would be known exactly and secondly, proportionality of amplitude and strength of disturbance implies continuous rather than discrete absorption. The model has to be modified by solving the Schrödinger equation for the harmonic oscillator [Eq. (5.3)].

Mechanical considerations are nevertheless useful, especially for the determination and visualisation of fundamental vibrations. In electronic spectroscopy, on the other hand, these models are useless; the reason for this difference is the relatively large mass of the nuclei involved in molecular vibrations compared to the electrons: dealing with molecular and especially with polymer vibrations we are a large step nearer to the macroscopic world.

5.3 Molecular Vibrations as Quantum Phenomena

Solving the Schrödinger equation for the harmonic oscillator [3] yields the following set of energy eigenvalues [Eq. (5.3)].

$$E = hv(v + 1/2); \quad v = 0, 1, 2 \dots . \tag{5.3}$$

The energy of the vibrational ground state ($v = 0$) is thus given by $E_0 = hv/2$.

The rest position at $x = 0$ in the mechanical model now changed into a maximum probability (ψ^2) at the same coordinate (see also Fig. 3.2). In the lower vibrationally excited states, ψ^2 (max) is not situated at the classical potential curve, only in the higher vibrational states the classical limit is approached in accordance with the "correspondence principle". Owing to the finite magnitude of the bond dissociation energy, Eq.

(5.3) is fulfilled only approximately and for small values of v, the region of vibrational spectroscopy; even here, $v(0\rightarrow2)$ is in general somewhat smaller than $2v(0\rightarrow1)$, whereas according to Eq. (5.3) the first "overtone" should have exactly twice the frequency of the fundamental. In higher excited states, the vibrational energy differences gradually diminish.

5.4 General Remarks on the Interpretation of Vibrational Spectra

Depending on the complexity of the systems studied and the level of accuracy attempted, two complementary approaches of interpretation have been developed[4, 7], one based on symmetry, force constants, selection rules etc. (molecular physics approach), the other one using rules derived from extensive comparative studies (empirical vibrational spectroscopy). This is shown schematically in Fig. 5.1.

In polymer spectroscopy, the molecular physics approach is largely restricted to crystalline polymers since in amorphous polymers and solutions the symmetry of the vibrating systems, the most important input parameter, is ill defined. The empirical vibrational spectroscopy is a good example of the "monomer aspect" of polymer spectroscopy discussed in Chap. 1. It is based on the fact that many vibrating subunits of

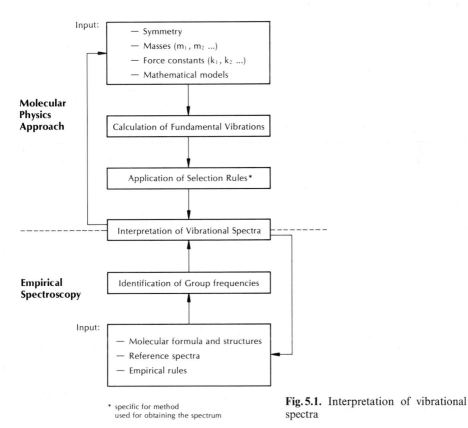

Fig. 5.1. Interpretation of vibrational spectra

* specific for method used for obtaining the spectrum

the polymer molecule are only weakly coupled with the rest of the molecule or with its surroundings so that the corresponding spectral signal always appears at about the same frequency. Small deviations from this frequency can be used to explain details of the mode of bonding of this group.

5.5 Symmetry and Fundamental Vibrations of One-dimensional Chain Molecules

Most of the characteristic features of ordered, linear polymers can be explained considering single chains. Additionally, splitting of peaks due to the (3D) crystal structure may occur provided that the unit cell contains more than one chain and the interaction between the non-equivalent chains is sufficiently large to be detected (factor group or Davydov splitting). The ideal polymer chain – in the simplest case the extended "zig-zag chain" – can be considered to be a 1D crystal. The prominent symmetry element of a 1D crystal is its translation (T) along the chain axis (see Fig. 5.2).

$$T_n(n = -\infty, \ldots -1, 0, +1, \ldots +\infty \text{"lattice constants" a})$$

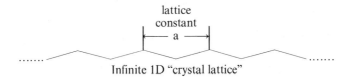

lattice
constant
a

Infinite 1D "crystal lattice"

Further symmetry elements of the 1D crystal lattice are [5, 6]:

C_n	Axis of symmetry
$\sigma(\equiv S_1)$	Plane of symmetry
i	Centre of symmetry
S_n	Rotation reflection axis of symmetry
$I(\equiv C_1 \equiv T_0)$	Identity.

In the following, the use of the symmetry elements will be discussed for extended chain polyethylene. It should be noted that the assumption of infinite chains seems to be justified even in case of folded chain crystals, due to the general "short sightedness" of spectroscopic methods (Chap. 1). An exception to this rule is the accordion vibration

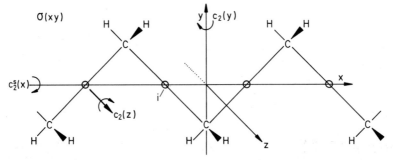

Fig. 5.2. Symmetry elements of the infinite extended chain conformation of PE. There is an infinite number of $C_2(y)$ and ($C_2(z)$) axes, inversion centres i and symmetry planes $\sigma(yz)$

detected in the Raman spectra of crystalline chain polymers, which is useful for determining the thickness of crystalline regions (Sect. 6.5) and thus depends on the finite length of the extended chain regions of the PE molecules.

The symmetry elements of the infinitely extended PE chain are indicated in Fig. 5.2, the plane of symmetry $\sigma(x, y)$ being the paper plane. The z-axis, therefore, is perpendicular to this plane, whereas x is the chain axis. All C-atoms are situated in the xy plane. Those symmetry operations which leave the system unchanged are called symmetry elements. The extended PE chain shown in Fig. 5.2 contains the following symmetry elements:

a) An infinite number of translations T_n.
b) Symmetry axes $C_2(y)$, $C_2(z)$; the x-axis is a screw axis ($C_2^s(x)$), i.e. rotation about $2\pi/n$ $(n=2)$ followed by T (1/2 lattice constant) leaves the molecule unchanged.
c) Symmetry planes $\sigma(xy)$, $\sigma(yz)$. The (xz) plane perpendicular to (xy) is a glide plane, since reflection about this plane has to be followed by T in order to leave the system unchanged.
d) Centre of symmetry: if the point i in Fig. 5.2 is taken as the origin of a Cartesian coordinate system, inversion (i.e. change of all signs) does not change the system.
e) Identity, I, visualised e.g. by rotation about $360\,°C$ $(2\pi/1)=C_1$.

These symmetry elements a) to e) form a space group [6)] and can be used for assignment of the (1D) crystal structure and for exactly describing the vibrations of the system. As already discussed, a vibration always constitutes a disturbance of the equilibrium. Consequently, the vibrating system can at most have equilibrium symmetry (totally symmetric vibrations). There is only a limited set of possible disturbances with regard to the individual symmetry elements which are codified for each (point) group in the form of character tables. The symbols of the "irreducible representations" i.e. allowed combinations of changes in the individual symmetry elements of a point group, are frequently used to classify the vibrations.

Space groups, which are used to describe the symmetry of crystals, can be written as a product of two subgroups $T \times G$, where T contains all translations and G, the factor group, contains all remaining symmetry elements and therefore describes the elementary cell of the crystal. If this contains only one molecule, G is identical with the point group of the molecule, otherwise G contains additionally the symmetry operations necessary for changing the molecules of the elementary cell which are not identical.

Since regular polymers have features both of molecules and crystals, it is possible to use space groups as well as point groups in order to describe their symmetry. Extracting the translational symmetry elements T_n in the case of the extended PE molecule, we obtain the elements of the factor group G:

$$I,$$

$$C_2(y); C_2(z); C_2^s(x),$$

$$i,$$

$$\sigma(xy); \sigma(yz); \sigma_g(xz).$$

The factor group G differs from the isomorphous point group D_{2h} related by calculation rules only by two translations of 1/2 lattice constant contained in $C_2^s(x)$ and $\sigma_g(xz)$ which do not affect the irreducible representations of D_{2h} shown in Table 5.1.

C. Vibrational Spectroscopy

Table 5.1. Character Table of the PE factor group, isomorphous with point groups D_{2h}

D_{2h}	I	$C_2(y)$	$C_2(z)$	$C_2^s(x)$	i	$\sigma(xy)$	$\sigma(yz)$	$\sigma_g(xz)$
A_g	1	1	1	1	1	1	1	1
B_{1g}	1	-1	-1	-1	1	1	-1	-1
B_{2g}	1	1	-1	-1	1	-1	-1	1
B_{3g}	1	-1	-1	1	1	-1	1	-1
A_u	1	1	1	1	-1	-1	-1	-1
B_{1u}	1	-1	1	-1	-1	-1	1	1
B_{2u}	1	1	-1	-1	-1	1	1	-1
B_{3u}	1	-1	-1	1	-1	1	-1	1

A: symmetric with regard to the main axis
g (gerade): symmetric with regard to inversion
u (ungerade): antisymmetric with regard to inversion
B: antisymmetric with regard to at least one C_2 axis
Assignment of axes see Fig. 5.2; notation of symmetry elements see text

The highest symmetry possible for the vibrating molecule (A_g) is the symmetry of the equilibrium. The corresponding wave function (ψ_0) retains its sign with regard to all symmetry operations. If the symmetry of the vibrating molecule is lower than equilibrium, one or several connected operations change the sign of the wave function, as shown in Eq. (5.4) for a reflection.

$$\Psi_v \xrightarrow{\ \sigma\ } (-1)\,\Psi_v. \tag{5.4}$$

The characters $+1$ and -1 in Table 5.2 indicate symmetric or antisymmetric behaviour of the wave function that describes the vibration with regard to the specific symmetry operation.

As can be seen from Table 5.2, the point group of the extended PE chain, D_{2h}, has eight symmetry elements and seven irreducible representations ($A_g \cdots B_{3u}$). These can be used to deduce the allowed vibrations, as described in detail by Tadokoro and Kobayashi [5] using matrix calculations. The coordinates describing the vibration are written as vectors (columns) and multiplied by quadratic matrices which describe the symmetry operations to be performed. In this way the new coordinates of the system can be found. For a complete theory of polymer vibrations, see Painter et al. [7]. Here, only a few of the fundamental vibrations obtained for PE will be shown.

One or more "symmetry coordinates" ($S_1, S_2\ldots$) can be ascribed to each irreducible representation. These contain the allowed changes of coordinates for all atoms of the unit cell. The totally symmetric representation A_g has the following symmetry coordinates (omitting normalisation factors of the order of 1):

$$S_1 = \Delta y(\text{C}_\text{I}) - \Delta y(\text{C}_\text{II}),$$
$$S_2 = \Delta z(\text{H}_\text{I}) - \Delta z(\text{H}_\text{II}) + \Delta z(\text{H}_\text{III}) - \Delta z(\text{H}_\text{IV}), \tag{5.5}$$
$$S_3 = \Delta y(\text{H}_\text{I}) - \Delta y(\text{H}_\text{II}) - \Delta y(\text{H}_\text{III}) - \Delta y(\text{H}_\text{IV}).$$

C_I and C_II are neighbouring carbon atoms of PE (Fig. 5.2), H_I and H_III are situated below the paper plane, H_II and H_IV above; the six atoms together form the unit cell of the

1D chain (rather than of 3D real PE crystals). S_1 describes the vibration $C_I\uparrow$ and $C_{II}\downarrow$, or vice versa. In S_2 the H atoms situated in the yz planes approach each other simultaneously. S_3 describes the concerted movement of $(H_I+H_{II})\uparrow$ and $(H_{III}+H_{IV})\downarrow$, or vice versa. It can easily be seen that all these vibrations are symmetric with regard to all symmetry operations of D_{2h}.

The irreducible representation A_u has only one symmetry coordinate (S_{10}) in accordance with symmetric behaviour with regard to the three C_2 axes and antisymmetry with regard to inversion and reflection at the three mirror planes (5.6).

$$S_{10} = \Delta x(H_I) - \Delta x(H_{II}) - \Delta x(H_{III}) + \Delta x(H_{IV}). \tag{5.6}$$

During this vibration, H_I and H_{III} approach each other in the chain direction (x), whereas H_{II} and H_{IV} move apart in this axis.

All in all, there are 14 fundamental vibrations of the (1D) unit cell [5]. Since the real orthorombic unit cell of PE crystals has a lower symmetry (C_{2h}), some symmetry restrictions may be eased in crystalline PE.

The calculation of the actual frequencies (v_v) is based on the masses, which are precisely known and on the valence forces and fields which are approximately known [7]. Together with isotopic substitution and measurements of band polarisations these calculations are useful in the identification of the spectral bands. Recently, a complete normal vibrational analysis of crystalline $PVCl_2$ has been reported [11] predicting Raman- as well as infrared absorption bands accurately. For sterical reasons many polymers are not able to crystallise in stretched form. These polymers often crystallise as helices which can also be treated as 1D systems [5]. In this case, the (1D) unit cell is given by one turn of the helix.

5.6 Phonons

In crystal lattices formed by atoms the only possible vibrations are those of the lattice "points". The number and frequencies of these collective vibrations depend on the strength of binding, the mass of the atoms and on the symmetry of the lattice. The lattice vibrations can be described by a wave and by a quantum (or particle) model, the quanta being called phonons [8]. A disturbance of the lattice, e.g. by sound or heat, propagates through the lattice in the form of these characteristic vibrations.

If the lattice points are occupied by molecules, complex ions, etc., intramolecular vibrations may occur in addition to the intermolecular or lattice vibrations. Since polymers can be regarded as molecules and 1D crystals, the term phonon can also be applied to the (mostly low-frequency) intermolecular vibrations and to intramolecular vibrations of the type discussed in Sect. 5.5.

A characteristic feature of the lattice waves or phonons is the dependence of their frequency on the phase difference (phase angle) of the vibration between neighbouring unit cells. This is shown schematically in Fig. 5.3 for a 1D lattice containing two atoms per unit cell. There are similar curves for longitudinal and transversal vibrations; for the purpose of illustration, however, the transversal picture has been preferred in Fig. 5.3. Since the dipole moment of the lattice can only change if the unequal (and differently charged) atoms move in the opposite direction during the vibration, only these vibrations are IR active and the corresponding (upper) branch of the dispersion relation

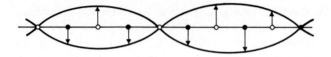

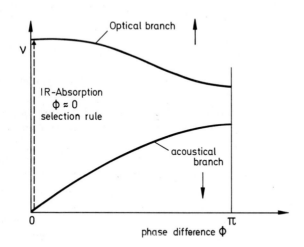

Fig. 5.3. Dispersion curve for longitudinal phonons in a 1D system. The illustrations on top (optical) and at the bottom (acoustical) are drawn for transversal vibrations for better demonstration of the basic difference of the two modes

is therefore called "optical branch". The lower branch is called "acoustical branch" owing to the low frequency propagation of acoustic waves which belong to this type of (synchronous periodic) movement of lattice points without change in dipole moments. It should be noted that the periodic change in density, corresponding to a similar change in polarisability, is favourable to the Smekal-Raman Effect (Chap. 6) of inelastic light scattering.

The absorption of IR photons is forbidden unless $\phi \approx 0$ since the phonon formed has to correspond not only to the IR energy (quantum resonance) but also to the wavelength of the exciting wave. The maximum phase difference therefore is given by Eq. (5.7).

$$\phi = \frac{2\pi a}{\lambda} \approx 0 \qquad (5.7)$$

a: lattice constant, λ: wavelength of IR radiation in the polymer.

In the case $\phi \approx 0$ indicated ($\lambda \gg a$) the corresponding particles in neighbouring unit cells are nearly all moving in the same direction. On the other hand, if $\lambda = 2a$; $\phi = \pi$, the corresponding particles in neighbouring cells move in opposite direction. The full exploration of the phonon dispersion, therefore, needs radiation of quantum energy corresponding to (far) IR photons, but wavelengths corresponding to X-rays, i.e. of the order of lattice constants; in other words, energy resonance and Bragg's diffraction of exciting radiation have to be fulfilled at the same time.

A radiation meeting these requirements is neutron radiation [10] of suitable particle velocity v. The relationship between kinetic energy ($mv^2/2$) and wavelength follows from Broglie's formula (5.8)

$$\lambda = \frac{h}{mv}. \tag{5.8}$$

The maximum energy which can be exchanged during collision of a particle wave in inelastic scattering therefore amounts to

$$E = \frac{h^2}{2m\lambda^2}. \tag{5.9}$$

For comparison, the analogous equation for electromagnetic radiations reads

$$E = h\nu = \frac{hc}{\lambda}. \tag{5.10}$$

The mechanism of inelastic neutron spectroscopy (Sect. 7.7.4) is momentum transfer between the particle wave (usually about $v = 800$ m s^{-1}, $E = 323$ J mol^{-1}, $\lambda = 0.5$ nm) and the polymer which is favoured if the masses of scattered particle and momentum-accepting atoms are of the same order of magnitude. The energy and momentum transfer occurs in an Anti-Stokes process from the polymer to the neutrons, thus increasing the velocity and decreasing the Broglie wavelength [Eq. (5.8)] of the neutrons. Both effects can be used for measuring INS. In order to get the full information needed for measuring the phonon dispersion curves, oriented polymers or crystals have to be used and the wavelength or velocity of the scattered neutrons has to be measured as a function of the scattering angle.

References

1. Zbinden, R.: Infrared Spectroscopy of High Polymers, New York: Academic Press 1964
2. Brügel, W.: Einführung in die Ultrarotspektroskopie, 4th ed., Darmstadt: Steinkopff 1969
3. Funck, E., Stuart, H.A.: Molekülstruktur, 3rd ed., Berlin, Heidelberg, New York: Springer 1967
4. Hummel, D.O. (ed.): Polymer Spectroscopy, Weinheim: Verlag Chemie 1974
5. Tadokoro, H., Kobayashi, M.: Vibrational Spectroscopy, in [4], p. 3
6. Hollas, J.M.: Symmetry in Molecules, London: Chapman and Hall 1972
7. Painter, P.C., Coleman, M.M., Koenig, J.L.: The Theory of Vibrational Spectroscopy and its Application to Polymeric Materials, New York: Wiley 1982
8. Kittel, C.: Introduction to Solid State Physics, 4th ed., New York: Wiley 1971. German translation, 3rd ed., München: Oldenbourg 1973
9. Allen, G., in: Ivin, J.J. (ed.): Structural Studies of Macromolecules by Spectroscopic Methods, London: Wiley 1976, p. 1
10. Allen, G.; Macromol. Chemie, Suppl. 3, 335 (1978)
11. Wu, M.S., Painter, P.C., Coleman, M.M.: J. Pol. Sci. Pol. Phys. ed. *18*, 111 (1980)

6 Raman Spectroscopy

6.1 Introduction

The vibrational spectroscopy of polymers comprises two important experimental methods:

- Infrared spectroscopy
- Raman spectroscopy.

The Raman spectroscopy of polymers, to be discussed in this Chapter uses visible laser radiation in the spectral region between 400 and 600 nm in order to excite polymer vibrations. The vibrational quanta $v'_v \approx 10$ to $4{,}000$ cm^{-1} are subtracted (in the more commonly studied case of Stokes lines) from the energy of the exciting photons and correspond to IR absorption between $\lambda = 1$ mm (fIR) to 2.5 µm (mIR/nIR). If light of $\lambda = 500$ nm ($v' = 20{,}000$ cm^{-1}) is used to excite the Raman spectrum, the above wave number range corresponds to the wavelength region from 500.25 to 625.00 nm in which the Raman spectrum is observed. Thus, Raman spectroscopy is experimentally a spectroscopic method operating in the VIS region (but not confined to it) and mechanistically belongs clearly to IR or vibrational spectroscopy (Chap. 5).

Raman spectroscopy is based on inelastic light scattering predicted theoretically by Smekal in 1923 [1, 2] and detected experimentally five years later by Raman [3, 4]. The inelastic light scattering is therefore called "Smekal-Raman Effect" [2], the spectroscopy based on this effect is generally called Raman spectroscopy.

In the field of Polymer spectroscopy, Raman spectroscopy has been applied after introducing laser excitation which removes some of the difficulties connected with strong (elastic) light scattering of polymers using the intense and highly monochromatic radiation which is typical of lasers [5, 19, 20]. In some early work, however, Raman spectra obtained using classical excitation sources were used in order to verify Staudingers theory of linear macromolecules [21, 22].

6.2 The Smekal-Raman Effect

If a clear non-luminescent liquid is illuminated laterally, a weak radiation of scattered light can be observed against a dark background. This Tyndall (or Rayleigh) scattering has the same frequency as the exciting radiation and its intensity (I_s) increases with increasing frequency or wave number according to Eq. (6.1).

$$I_s \sim v'^n.$$

$$1 \leqslant n \leqslant 4.$$

(6.1)

The Tyndall effect is due to density variations in the scattering medium and can be observed in gases (blue sky) and solids as well as in liquids. It should be minimal in an ideal crystal near O K. The origin of this scattering is visualised by assuming the electrons to perform vibrations in phase with the exciting radiation (but without "true" absorption). The vibrating electrons are thought to be the origin of new waves which in the case of ideal order all vanish by interference, except those which are in phase with the original wave (i.e. no scattering in this case). If this order is disturbed, however, quenching by interference is incomplete thus giving rise to scattering. In polymer research, this type of (elastic) scattering is used for measuring the size ($\bar{M}_w$) and shape of macromolecules.

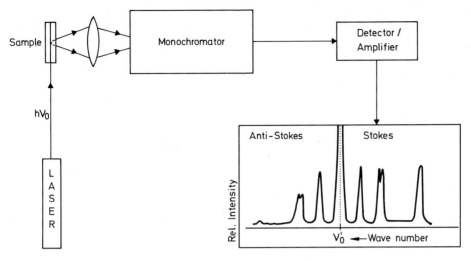

Fig. 6.1. Schematic view of a laser Raman spectrometer and resulting spectrum

If monochromatic light is used for excitation (see Fig. 6.1) and the radiation scattered by the sample is analysed by means of a monochromator, in addition to the relatively strong Tyndall component (v_0') additional weak peaks appear both at lower (Stokes, $1, 2 \ldots n$) and higher (anti-Stokes, $-1, -2 \ldots -n$) wave numbers compared to v_0'. These components indicating the Smekal-Raman effect are characterised by the following features [2, 4]:

a) The difference in wave numbers $\Delta v' = v_0' - v_n$ and $\Delta v' = v'_{-n} - v_0'$ is independent of the frequency of excitation (v_0') and independent of the direction of observation.
b) The intensity of the anti-Stokes lines decreases in a Boltzmann relationship with increasing $v'_{-n} - v_0'$.
c) The inelastically scattered radiation is characteristic of the scattering substance with regard to $\Delta v'$ and relative intensity of the lines.
d) The individual peaks partly correspond to infrared lines ($\Delta v' = v_v'$), but not so in highly symmetric molecules.
e) The intensity is roughly proportional to the density of the medium, in contrast to Tyndall scattering which is strong in gases due to the pronounced density fluctuations in this state of aggregation.
f) If linearly polarised light is used for exciting, the Smekal-Raman lines are likewise polarised, but different lines to a different extent.
g) Intensity increases strongly with increasing frequency, similar to Tyndall scattering [Eq. (6.1)].

All effects observed can be explained by two assumptions:
- The spectrally shifted lines are due to the transfer of vibrational quanta to the scattering medium (v_n') or by the acceptance of vibrational quanta from the medium (v'_{-n}).
- The selection rule is the change in polarisability during the vibration, favouring the IR-inactive totally symmetric vibrations. (Rotations will not be regarded here, since they are not relevant to polymers.)

C. Vibrational Spectroscopy

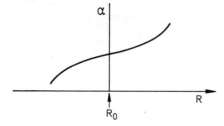

Fig. 6.2. Change in polarisability as a function of vibration coordinate (schematic) for a totally symmetric vibration

Examples of totally symmetric vibrations in small molecules are

the "breathing" vibrations (C_6H_6; CCl_4) and symmetric stretching vibrations (CO_2) in which the dipole moment remains zero $\frac{d\mu}{dR}=0$ during vibrations, hence IR inactive ($R=$coordinate of vibration).

The polarisability (α), however, changes in this type of vibration (Fig. 6.2)

$$\frac{d\alpha}{dR}\neq 0$$

near the equilibrium position.

This can be understood considering the strong change in electron density during these vibrations. The polarisability of a molecule is a measure of the ease of electron displacement under the influence of an electric field (ε) inducing a dipole moment μ_{ind}, Eq. (6.2)

$$|\mu_{ind}|=\alpha|E| . \tag{6.2}$$

Except for molecules with spherical symmetry, the induced dipole moment depends on the orientation of the molecule relative to the electric field. The polarisability, therefore, is a tensor which has a maximum of six distinct, real components ($a_{xx}, a_{yy}, a_{zz}, a_{xy}, a_{xz}, a_{yz}$), if a suitable Cartesian coordinate system is chosen for each molecule.

In symmetric molecules, some of the components of the polarisability tensor coincide. The basic difference between direct absorption of electromagnetic radiation [Eq. (3.3)] and the Smekal-Raman-Effect consists in the replacement of the dipole moment operator by the polarisability operator leading to an induced rather than a "real" transition dipole moment [6].

$$M_{k,n}(ind)= <\varphi_k|\alpha|\varphi_n>E_0 . \tag{6.3}$$

Due to the large wavelength of the exciting light, compared to the size of the absorbing groups, the electric field vector can be considered constant and thus written outside the

64

integral (6.3). The polarisability tensor acts as operator on the (vibrational) wave functions of the initial (k) and final (n) state of the system and hence Raman absorption can only be observed if there is a significant change in polarisability in moving from φ_k to φ_n.

Quanta may be transferred between the radiation field and the vibrating system in both directions leading with the same probability to Stokes lines and anti-Stokes lines. However, the density of vibrating molecules that are able to cause anti-Stokes lines depend on the vibration energy $h\Delta v$ and the temperature according to Boltzmann (6.4).

$$\frac{N'}{N} \approx e^{-\frac{h\Delta v}{kT}}.$$ (6.4)

The anti-Stokes lines of high energy vibrations, therefore, are much weaker than the corresponding Stokes lines. The intensity ratio furthermore depends on the frequency v_0, since the intensity of Raman scattering increases with frequency in the same way as does Tyndall scattering (6.5).

$$\frac{I \text{ (anti-Stokes)}}{I \text{ (Stokes)}} = \frac{v_{-n}^4}{v_n^4} \cdot e^{-\frac{h\Delta v}{kT}}.$$ (6.5)

Polarisation of Raman lines is strong if the vibration and the molecule are symmetric and thus can be used to identify specific vibrations of a molecule. Due to strong (elastic) light scattering, however, polarisation measurements have rarely been performed on polymer samples.

Summing up we can state that the Smekal-Raman-Effect is a light scattering effect rather than an absorption process. Although vibrational quanta are taken up or released by the molecules, the role of the exciting light is to create induced dipoles not corresponding to any real energy levels of the molecule. The selection rules are characteristic of scattering and not of the absorption and emission of photons.

6.3 Experimental

The possibility of applying Raman spectroscopy to polymer problems is largely due to the development of laser excitation. A schematic view of a Raman spectrometer has already been given in Fig. 6.1.

Polymers tend to strong Tyndall scattering, especially in the partly crystalline state (spherulites) and dissolved in solvents (polymer coils). The exciting light, therefore, has to be highly monochromatic, so that v_0' can be separated from v_n and v_{-n}'. Even if lasers are used ($\Delta \approx 1/3 \text{ cm}^{-1}$), double or triple monochromators have to be employed. The coherent radiation of lasers offers the additional advantage of easy focusing so that very small samples can be investigated. On the other hand, large samples are not advantageous so that the analytical value of Raman spectroscopy with regard to the detection of impurities, small concentrations of additives, etc. is rather low.

The intensity of the Raman spectrum is of the order of 10^{-9} of the exciting light. Thus, a high performance monochromator is required. "Holographic" gratings [7], as produced by Jobin-Yvon in France, seem to allow the use of a single monochromator, if highest resolution is not necessary. According to Bulkin [7], the combined effect of intense laser radiation + holographic gratings + red sensitive photomultipliers has increased the sensitivity of Raman spectroscopy 100 times within ten years.

Choosing the optimum excitation frequency involves a compromise between several conflicting requirements:

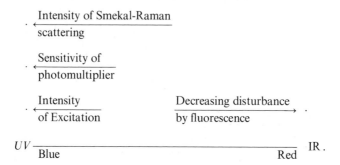

Since the fluorescence of the sample excited by v'_0 seems to constitute the greatest experimental problem in polymers, UV excitation is generally avoided, although all other factors call for short wavelength excitation.

Fluorescent groups, if caused by impurities (Sect. 4.4.2), are often removed by "burning out", i.e. by photochemical degradation. Red-sensitive, cooled photomultipliers and photon-counting devices are also useful in preventing any excessive disturbance by fluorescence.

As a compromise, excitation by Ar^+ lasers (0.2 to 1 W) near 500 nm (488.0 and 514.5 nm) is often used. In the red, He/Ne lasers (632.8 nm) and Kr^+ lasers (647.1 nm) can be used.

The wave number difference $\Delta v' = v_0 - v'_n$ is recorded in the Stokes region of the spectrum so that spectra recorded with different v'_0 can be compared with each other and with IR absorption spectra in wave number presentation.

The disturbance caused by fluorescence of the sample can be reduced by many orders of magnitudes in CARS (coherent anti-Stokes Raman scattering) spectroscopy. In this technique two laser beams are crossed at the sample with a small angle, one of the laser beams being tunable in frequency. Strongly enhanced scattering is observed if the difference of the two laser frequencies corresponds to a polymer vibration [7, 20].

The coherently scattered radiation is to the short-wavelength side of fluorescence and spatially directed; it can therefore be separated from fluorescence.

6.4 Examples of Laser-Raman Spectra of Synthetic Polymers

Raman spectroscopy has been used mainly to answer questions relating to the physical structure rather than to analytical-chemical aplications. Therefore, chemically simple polymers, such as PE, PTFE, PP [8-10] and linear paraffins as oligomeric models [9, 14] have been studied preferentially. In these linear polymers, differences between crystalline and amorphous and solid/molten state can be observed most easily. The wave numbers of Raman active 1D chain vibrations of PE are summarised in Table 6.1.

Table 6.1. Raman-active vibrations in PE

$\Delta v'$ [a] cm^{-1}	Interpretation [5]	(1D) Symmetry [b]
2,850	C—H Stretching vibration	A_g
2,884	[c]	?
2,920	C—H Stretching vibration	B_{1g}
1,060	C—C (Skeleton) stretching vibration	B_{2g}
1,130	Skeleton	A_g
1,170	CH$_2$ Rocking	B_{1g}
1,300	CH$_2$ Twisting	B_{3g}
1,420	Overtone	
1,440	CH$_2$ Deformation	A_g

[a] Data (rounded) according to Schaufele [9] and Strobel [11]

[b] See Sect. 5.5

[c] Very intense band [5]; there are several more Raman lines in this frequency region

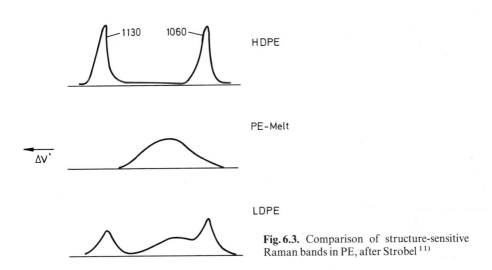

Fig. 6.3. Comparison of structure-sensitive Raman bands in PE, after Strobel [11]

A good example of sensitivity of structural order is given by the 1,130/1,060 lattice vibration of PE [11], see Fig. 6.3.

Other crystallinity/conformation peaks show an analogous behaviour. Certain bands are associated with a special type of crystal lattice and therefore have to be ascribed to an 3D rather than a 1D structure. In the amorphous and molten state, the intensity of interactions between molecules fluctuates randomly thus causing broadening of the spectra. In the crystalline phase, on the other hand, each molecule "feels" the same local environment so that the peaks are narrower because they are more homogenous. A similar effect is exerted by low temperatures where additionally bands originating from vibrationally excited groups ("hot bands") are suppressed.

Splitting of bands occurs in crystallites if more than one chain is found per unit cell, provided the interaction (coupling) is strong enough so that splitting can be measured.

C. Vibrational Spectroscopy

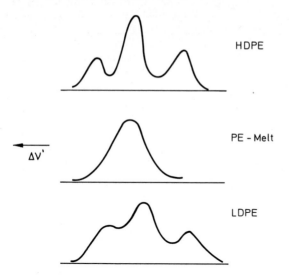

HDPE

PE - Melt

$\Delta v'$

LDPE

Fig. 6.4. Fermi splitting in the deformation region of PE

In the simplest case, the non-equivalent chains can vibrate in phase with each other or in the opposite sense, e.g. in orthorhombic PE containing two chains per unit cell. In order to observe splitting caused by the two different vibrations experimentally, the strength of the interaction, as given by the splitting of Raman frequencies ($\delta \Delta v'$), has to be stronger than the spectral resolution. At 77K, the Raman spectrum of crystalline PE indeed shows two splittings which can be attributed to 3D crystal structure;

$$\Delta v' \ (cm^{-1})$$

$$C-C \quad 1,063 \quad \rightarrow \quad \begin{matrix} 1,062 \\ 1,064 \end{matrix} \quad \delta \Delta v' = 2 \ cm^{-1}.$$

Stretching

$$CH_2 \quad 1,295.5 \rightarrow \begin{matrix} 1,293 \\ 1,295.5 \end{matrix} \quad \delta \Delta v' = 2.5 \ cm^{-1}.$$

Twisting

As can be seen, the interaction energy between the two non-equivalent chains is quite small; comparable Davydov-splittings of allowed electronic transitions are in the range of 100 to 1,000 cm^{-1}. Linear paraffins which can be considered as oligomeric models of linear PE crystallise in an orthorhombic crystal lattice comparable to PE and additionally in a triclinic modification which contains only one chain per unit cell. In contrast to the orthorhombic structure which behaves analogous to PE, the latter does not show splitting.

A further effect which may increase or superpose the above splitting is due to Fermi resonance. An example of this splitting is given in Fig. 6.4. Fermi resonance is due to the coincidence of a weak overtone or combination vibration with a strong Raman-allowed band of the same symmetry. The strong transition transfers part of its strength to the weak one and instead of one strong peak (hiding the weak overtone) we observe a strong couple of bands.

The few experimental examples given show that the spectrum of the partly crystallised polymers is roughly a superposition of a true crystal spectrum and a melt spec-

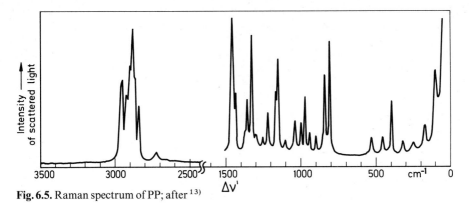

Fig. 6.5. Raman spectrum of PP; after [13]

Table 6.2. Raman-active vibrations in PTFE [5]

$\Delta v'$ cm^{-1} [a]	Interpretation
1,381	Symmetric stretching (CF_2)
732	vC—C skeleton vibration
383	CF_2 deformation
290	CF_2 twisting

[a] Only strong lines indicated

trum. In principle, this offers the possibility of measuring the degree of crystallinity (in addition to X-ray and density measurements). However, comparing different methods of measuring the degree of crystallinity in polymers, the physical effect used has to be checked with regard to the minimum crystallite size which can be detected. X-ray diffraction, e.g., requires perfect order over a distance of at least 5 nm to ensure sharp reflexes; this corresponds to a minimum crystallite volume of about 100 nm^3. Raman-spectroscopically, according to Hendra [5], the minimum length of a 1D chain contains 5 to 6 monomer units and 8 to 10 chains laterally are required if 3D crystal effects are to be observed. As an order of magnitude, therefore, the minimum crystallite volume amounts to only 1 to 10 nm^3. Frequently, characteristic 1D (conformation) bands are used to estimate the degree of crystallinity, but this does not necessarily indicate the true 3D crystal content.

Isotactic poly(propene) (Fig. 6.5) crystallises in the conformation of a 3/1 helix (3 monomers per turn). The 3D crystal unit cell contains four chains whose interaction seems to be too weak in order to show any detectable Davydov splitting (as discussed for PE, see above). It is interesting to note that the helix bands do not vanish during melting when the 3D order is lost. It is only 10 K above the melting point that the melt shows the spectrum characteristic of the amorphous polymer, also shown by solid, atactic PP. Melting, therefore, leads to the loss of long-range order, whereas the 1D order persists, provided the polymer still forms helices (the PE crystal spectrum conversely collapses at the melting point).

Polytetrafluoroethene (PTFE) at room temperature shows one chain per unit cell (no Davydov splitting to be expected), wound up in the form of an elongated (15/1)

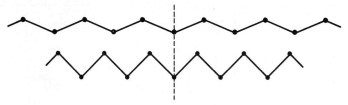

Fig. 6.6. Fundamental vibration of the longitudinal accoustic vibration of a linear zig-zag chain (accordion vibration)

helix. The prominent Raman peaks are compiled in Table 6.2. Owing to the higher mass of F atoms, the CF_2 stretching vibration is at much lower frequency compared to CH_2; the symmetric component is preferred in the Raman spectrum, whereas the opposite is the case in IR absorption. Doubling of several peaks at low temperature indicates a second crystal structure with two chains per unit cell. The Raman spectra of several polymers have been discussed and compared with normal coordinate analysis by Painter et al. in Chap. 16 of their book [15].

6.5 The Accordion Vibration

Among all Raman-active polymer vibrations, the accordion vibration has found the widest application. In the fundamental vibration, all C—C—C angles change simultaneously so that a motion described graphically by the name of this vibration results. The vibration is totally symmetric (A_g) and thus Raman-active. The same applies to all uneven overtones.

This family of vibrations is also named longitudinal accoustic vibrations owing to the similarity with accoustic waves (see discussion of phonons, Sect. 5.6). In order to describe the accordion vibration quantitatively, an equation similar to that for a vibrating spring can be used [5] in a first approximation:

$$v = \frac{m}{2L} \sqrt{\frac{E_c}{\varrho}} \quad (m = 1, 2, 3 \dots). \tag{6.6}$$

v = Frequency = Δv in Raman spectroscopy
L = Length of the vibrating molecule or part of molecule
$\dfrac{2L}{m}$ = Wavelength of (acoustic) vibration ($m = 1 : \lambda = 2L$)
E_c = Elastic constant (Young's Modulus)
ϱ = Density

If this equation is to be applied to polymers, e.g. crystallising in the form of extended or folded chains, the elastic constant and density of these regions has to be used instead of the (averaged) macroscopic properties.

Crystallised paraffins have been used in order to calibrate the Raman bands, up to $n = 100$ [9], see Table 6.3. From these data and the experimental lamella thickness, $E_c = 3.6 \times 10^{11}$ Pa has been found for these model compounds.

Several simple polymers crystallise in the form of lamellae as well; this is shown schematically in Fig. 6.7.

The lamellae consist of a highly ordered crystalline core of thickness L_c and a less ordered folding layer, often considered to be amorphous. Since $L_c \approx 10$ nm, the ac-

Table 6.3. Accordion vibrations in paraffins as oligomeric models of linear PE [a]

Paraffin	Order of vibration m	$\Delta v'$ (cm^{-1})
$C_{20}H_{42}$	1	114
	3	324
	5	475
$C_{32}H_{66}$	1	76
	3	211
	5	337
$C_{94}H_{190}$	1	26
	3	71
	5	121
	.	
	.	
	25	491
	.	
	.	
	31	556

[a] Data according to [9]

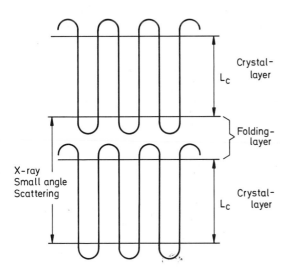

Fig. 6.7. Lamellar structure of folded chain polymer crystals (schematic)

cordion fundamental vibration is expected to be at low frequency [experimentally observed at 10 to 30 cm^{-1} ($m = 1$)] and the $m = 3$ overtone at about 3 times this frequency. For simple estimates of lamella thickness in polymers, the paraffin results can be used as a preliminary calibration. A more detailed analysis by Strobl[11], however, shows that coupling between the lamellae has to be introduced into the calculation; this results in an increase in frequency compared to Eq. (6.6). A refined value of $E_c = 2.9 \times 10^{11}$ Pa has been determined for PE (HDPE), yielding Eq. (6.7).

$$\Delta v'_{(m=1)} = 2.8 \times 10^{-5}/L. \tag{6.7}$$

Comparing L (Raman) obtained according to Eq. (6.7) with L_C, as obtained by chemical edging and L_X (X-ray small-angle scattering) we find:

$$L \approx L_X > L_C$$

e.g. for solvent-crystallised linear PE ($\bar{M} \approx 2 \times 10^4$):

Raman: $L = 11.7$ nm
$\quad\quad\quad L_X = 12.0$ nm
$\quad\quad\quad L_C = 9.6$ nm.

These data indicate that the folding layer is more anisotropic (less amorphous) than believed so far.

Other polymers investigated include [5]:
- (cis)poly(1,4-butadiene)
- Poly(tetrahydrofuran) and other poly-ethers
- Oligo-TFE
- Oligo-ethylene oxide

Raman studies using the accordion vibration as a probe involve studies on the growth of lamellae, crystallisation, melting, etc. Since the laser beam can easily be collimated (diameter ≈ 100 μm), even local changes of lamellar structure can be resolved and measured.

6.6 Resonance Raman Scattering

If the frequency of the exciting radiation corresponds, or almost corresponds, to an electronic absorption band the Raman lines due to vibrations coupling with the electronic transition are much stronger than ordinary lines. This effect is called "resonance Raman scattering" and can not in general be observed if the sample shows strong fluorescence (Fig. 6.8). The influence of fluorescence which, in contrast to Raman scattering, is a delayed emission (Chap. 4), can be minimised by fluorescence quenching or choosing v_0' just below the O—O level of the transition to be studied. The main differences between resonance Raman scattering and ordinary Raman scattering are:
- The effect is observed only near electronic absorption bands.
- It is much stronger and thus more easily detected.

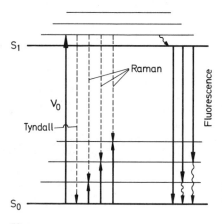

Fig. 6.8. Resonance Raman scattering and fluorescence (schematic)

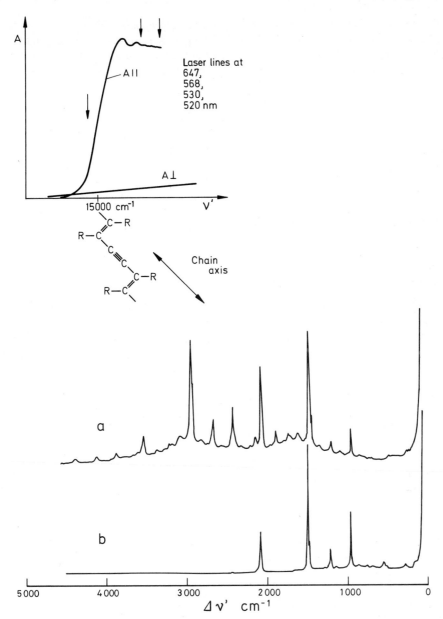

Fig. 6.9 and 6.10. Absorption (6.9) and resonance Raman scattering (6.10) of single crystals of PTS, redrawn after Bloor et al. [12]. Excitation in Fig. 6.10 at $\lambda_{exc} = 568.2$ nm (*a*) and 647.1 nm (*b*)

– Only such vibrations show increased intensity which are coupled with the electronic transition and therefore can be distinguished from other vibrations not belonging to the chromophore.

An experimental example of an application of resonance Raman scattering to polymer problems consists in the study of conjugated poly-diine single crystals [12, 16, 17]. As can be seen from Fig. 6.9, the strong electronic absorption near 15,000 cm^{-1} is polarised along the polyconjugated "backbone" of the polymer. In the resonance Raman spectrum, therefore, characteristic vibrations of the backbone are to be expected, whereas the aromatic side groups should not contribute to the spectrum (Fig. 6.10 b). Strong lines at 1,485 cm^{-1} (st C=C), 2,086 cm^{-1} (st C≡C), 1,203 cm^{-1} (b C=C) and 953 cm^{-1} (b C≡C) prove the existence of double and triple carbon bonds in this polymer and at the same time their vibrational coupling with the electronic transition.

The strong resonance Raman lines associated with the conjugated backbone has been used in order to study the solid-state polymerisation of single crystals of the monomers 1,6-di-p-Toluene sulfonyloxy-2,4-hexadiine (TSHD)[16] and the corresponding p-methoxybenzene analogue[17]. The interest in this work, as well as in ESR studies to be discussed in Sect. 9.7 lies in a detailed understanding of the polymerisation leading to polyconjugated single crystals, as first described by Wegner[18]. The Raman spectra of PTS are shown to be sensitive to the mechanical strain exerted on the polymer chain during polymerisation[15] due to the change in lattice dimensions.

6.7 Résumé of Raman Spectroscopy of Polymers

Table 6.4 shows an impressive multitude of phenomena which can be studied by Raman spectroscopy. In contrast ot IR spectroscopy, it can be applied to aqueous solutions. This point is attractive for studies of biopolymers whose vibrational spectra consequently can be studied in their natural environment. Conformation and crystallinity directly influence the Raman spectra of polymers, whereas tacticity has an indirect influence by favouring certain conformations and crystal structures. The electronic structure influences the binding strength of the vibrating groups and thus the spectral position of the corresponding bands; in resonance Raman scattering, the electronic structure contributes directly to Raman scattering due to selective coupling of the chromophore with its characteristic vibrations and corresponding selectivity of the enhanced scattering. Movements of the chains and crystal lattice (phonons), filtered by the selection rule of polarisability change are the very reason of Raman scattering. Complex formation, finally, can likewise be detected in the Raman spectra, e.g. H-bonds.

For experimental purposes the small sample volume turns out to be advantageous. Furthermore, the whole frequency range can be recorded using one spectrometer. In the development of Laser Raman spectrometers progress has been made by combining intense laser sources (variable v_0), monochromators involving only minimal losses of

Table 6.4. Informations obtained by Raman spectroscopy relating to structure and dynamics of polymeric systems

Structure		Dynamics	
Chemical structure	+	Movements of the chain,	+
Tacticity	(+)	segments and side groups	
Conformation	+	Phonons	+
Crystallinity	+	Excitons	−
Electronic structure	+	Complex formation and related phenomena	+

light due to scattering, red sensitive photomultipliers, photon counting and use of computers. An increased use of polarised Raman spectra, resonance Raman scattering and CARS is expected to further extend the scope of Raman spectroscopy of polymers which so far has turned out to be of prime importance to:

- analysis of polymer vibrations (complementary to IR absorption)
- conformation studies
- polymer lamellae (micromorphology)
- vibrational spectra of water-soluble polymers, especially biopolymers
- polyconjugated single crystals.

References

1. Smekal, A.: Naturwissenschaften *11*, 873 (1923)
2. Kohlrausch, K.W.F.: Der Smekal-Raman-Effekt, Berlin: Springer 1931
3. Raman, C.V., Krishnan, K.S.: Nature *121*, 501 (1928)
4. Long, D.A.: Raman Spectroscopy, New York: McGraw-Hill 1977
5. Hendra, D. in: Hummel, D.O. (ed.): Polymer Spectroscopy, Weinheim: Verlag Chemie 1974, p. 151
6. Herzberg, G.: Molecular Spectra and Molecular Structure. I. Spectra of Diatomic Molecules, 2nd ed. New York: Van Nostrand 1950
7. Bulkin, B.J. in: Brame, E.G., Jr. (ed.): Applications of Polymer Spectroscopy, New York: Academic Press 1978, p. 121
8. Hendra, P.J. in: Adv. Pol. Sci. *6*, 151 (1969)
9. Schaufele, R.F.: Macromol. Rev. *4*, 67 (1970)
10. Andrews, R.D., Hart, T.R. in: L.H. Lee (ed.): Characterization of Metal and Polymer Surfaces, Vol. 2, New York: Academic Press 1977, p. 207
11. Strobel, G.R.: Colloid Polym. Sci. *257*, 584 (1979)
12. Bloor, D., Preston, F.H., Ando, D.J., Batchelder, D.N., in: K.J. Ivin (ed.): Structural Studies of Macromolecules by Spectroscopic Methods, London: Wiley 1976, p. 91
13. Schrader, B., Meier, W. (eds.): DMS (Raman/IR) Atlas, Weinheim: Verlag Chemie 1974/75, Vol. 1 + 2; Polymers: Sect. N, Vol. 2
14. Zerbi, G., Magni, R., Gussoni, M., Holland Moritz, K., Bigotto, A., Dirlikov, S.: J. Chem. Phys. *75*, 3175 (1981)
15. Painter, P.C., Coleman, M.M., Koenig, J.L.: The Theory of Vibrational Spectroscopy and its Application to Polymeric Materials, New York: Wiley 1982, p. 377
16. Bloor, D., Kennedy, R.J., Batchelder, D.N.: J. Pol. Sci. Pol. Phys. Ed. *17*, 1355 (1979)
17. Bloor, D., Ando, D.J., Hubble, C.L., Williams, R.L.: J. Pol. Sci. Pol. Phys. Ed. *18*, 779 (1980)
18. Wegner, G.: Z. Naturforsch. *24* b, 824 (1969)
19. Cutler, D.J., Hendra, P.J., Fraser, G., in: Dowkins, P.V. (ed.): Developments in Polymer Characterisation, Vol. 2, Applied Science Publ. (1980), Chap. 3
20. Lascombe, J., Huong, P.V. (eds.): Raman Spectroscopy, Linear and Nonlinear, Proceedings of the Eighth International Conference on Raman Spectroscopy, Bordeaux, France, Wiley Heyden, Chichester (London) 1982
21. Signer, R., Weiler, J.: Helv. Chim. Acta *15*, 649 (1932)
22. Mitushima, S.-I., Morino, Y., Inoue, Y.: Bull. Chem. Soc. Japan *12*, 136 (1937)

7 Infrared Spectroscopy of Polymers

7.1 Introduction

Historically, studies of the selective absorption of infrared (IR) radiation preceeded vibrational spectroscopy using the Smekal-Raman effect, although in the early times recording of IR-spectra was difficult and time-consuming[1-4]. The development of

C. Vibrational Spectroscopy

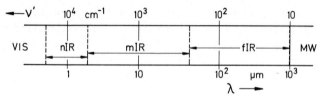

Fig. 7.1. The IR region of the spectrum of electromagnetic radiation

routine double beam spectrophotometers operating in the analytically most important medium IR (mIR) made this type of spectroscopy the most important one in identifying known compounds and elucidating the chemical structure of unknown ones.

Since the techniques of IR-absorption spectroscopy are well suited for polymers, a large number of polymer spectra has been collected and published [5]. Spectra of polymer films are even used to calibrate the IR-spectrophotometer. It is usual to divide the IR-region into three parts (Fig. 7.1):

$$\text{nIR: } 0.75\text{--}2.5\ \mu\text{m } (13{,}300\text{--}4{,}000\ \text{cm}^{-1})$$
$$\text{mIR: } 2.5\text{--}50\ \mu\text{m } (4{,}000\text{--}200\ \text{cm}^{-1})$$
$$\text{fIR: } 50\text{--}1{,}000\ \mu\text{m } (200\text{--}10\ \text{cm}^{-1}).$$

In the nIR [6] we find the weak overtone ($v = 2, 3, \ldots$) and combination bands whose fundamentals, mostly stretching vibrations, are in the mIR. The nIR requires either concentrated solutions or a film thickness of at least 0.1 mm. Experimentally, the nIR is covered by many UV/VIS-spectrophotometers ("nIR-bonus") using the tungsten lamp as radiation source, quarz optics and a PbS detector. The resolution is good and several peaks which can be distinguished in the $\Delta v = 1$ transition only with difficulty are separated at $\Delta v = 2$, roughly at $2 \times v'_0$ of the ground vibration. There are only few papers on polymer spectroscopy in the nIR [8], but a closer look at this neglected field would be worthwhile. The nIR identification table by Goddo and Delker [9] is also reproduced in [3].

The nIR spectra have been used in order to evaluate the structural order (crystallinity) and quantitative composition of polymers (additivity of C—H increments of optical density) [8].

In addition to these higher excited vibrational states, there are few long-wavelength electronic bands extending into the nIR, e.g. those of some dyes, strong CT complexes, radical ions and extensively conjugated double bonds. These absorptions are in general very strong and may be observed, e.g. in coloured, degraded and electrically conducting polymers.

7.2 Absorption of Infrared Radiation

The IR-spectroscopy of polymers is based on excitation of polymer vibrations (Chap. 5) by absorption of photons in the spectral regions indicated in Sect. 7.1. The most important selection rule is the change in the dipole moment during excitation $d\mu/dt$, the same as in electronic excitation (UV/VIS). This selection rule is to be attributed to interaction of the polymer with the electric field component of the electromagnetic radiation. Totally symmetric vibrations – highly active in Raman spectroscopy – are forbidden in IR absorption, since by definition the dipole moment cannot change during such a transition (examples see Sect. 6.2). Strong IR-absorptions are displayed by polar groups which already in the ground state have a strong dipole moment (μ_0), perhaps the best known example being the carbonyl group that is present in many polymers as part of the basic unit (polyesters, polyamides) or as artefact:

$$(+) \quad (-)$$
$$>C=O \quad \mu_0 = 2 \text{ to } 3 \text{ Debye}.$$

Vibrations of this kind are strongly localised in the "chromophore" and therefore invariably are found in a rather narrow spectral region. Small, but analytically significant shifts are due to neighbouring groups, in the example of $C=O$ groups, e.g., conjugation with $C=C$ double bonds which alters the binding strength and thus the resonance frequency according to Eq. (5.1). The intensity of the carbonyl band amounts to

$$\varepsilon_{max} \approx 100 \text{ to } 500 \text{ l mol}^{-1} \text{ cm}^{-1}.$$

The quantitative description of an IR peak requires at least the spectral position (λ or, better, v') and a measure of intensity, e.g. the molar decadic absorption coefficient (ε) of the absorption maximum.

$$\varepsilon = \frac{OD}{c \cdot d} \quad (1 \text{ mol}^{-1} \text{ cm}^{-1}) \tag{7.1}$$

OD = optical density (absorbance) at a specific wavelength
c = molar concentration given in mol l^{-1}
d = thickness of film (solid state spectrum) or cuvette (solution) given in cm.

Equation (7.1) is another form of Lambert-Beer's law [Sect. 3.2.3, Eq. (3.8)] whose validity in IR-absorption has been questioned. Especially OH and NH peaks show deviations from simple behaviour. It has been suggested [7] to use the area of the absorption band ($\int OD(v')dv'$) rather than OD itself in (7.1) in order to compensate for line broading effects which may be due to association etc. In practice, unfortunately, only qualitative statements such as vw, w, m, s, vs. are made in most papers reporting mIR spectra of polymers as well as monomers which are mostly plotted in "percent transmission" (100 T) as a function of λ or v'. It is usually not possible to calculate absorption coefficients from published spectra because data on d and/or c as needed for Eq. (7.1) are missing. These data are indeed difficult to obtain in case of meltpressed films or K Br pills, but should always be reported if homogeneous films or solutions are investigated. For the sake of comparison with electronic absorption bands, it should be noted that the range of useful mIR absorption intensity is roughly given by

$$\varepsilon_{max} \approx 10 \text{ to } 500 \text{ l mol}^{-1} \text{ cm}^{-1};$$

this corresponds to the intensity of symmetry-forbidden UV-absorption bands. Absorptions in the fIR are often found to be weaker than mIR bands. They frequently are due to excitation of phonons, i.e. lattice vibrations (Sect. 5.6).

C. Vibrational Spectroscopy

For analytical purposes, relative intensities of different peaks of the spectrum may be used after calibration with suitable reference compounds, e.g. a series of homologous molecules.

7.3 Experimental

7.3.1 IR-Absorption

Two types of IR spectrometers are at present available and suited for recording polymer spectra;
– dispersive double beam spectrometers
– computerised Fourier Transform Interferometers (FTIR) [10, 11].

In their measuring principle double beam spectrometers [3] using a dispersive element (prism or grating) are similar to UV/VIS-spectrometers. Continuous IR radiation produced by a heated (rare earth oxide) rod is divided into two parallel beams which pass sample and reference, the monochromator and are finally detected by a suitable IR detector and recorded as the intensity ratio of the two beams $I/I_0 (= T)$. The optical components are reflecting mirrors and gratings or prisms made from non-absorbing material, e.g. NaCl for the range 2 to 15 µm.

Film-forming polymers can be conveniently measured in form of thin films. Strong disturbances of the spectra may result from periodic interference patterns which are superposed for the true absorption or transmission spectrum. This disturbing effect can be prevented by casting the polymer solution on to a rough surface [8], followed by slow evaporation of the solvent. The optimal thickness amounts to $d = 10$ (polar polymers) to 30 µm (non polar polymers).

Dispersive IR spectrometers are less suited for the fIR range owing to the extremely weak intensity of thermal radiation sources, strong absorption of water traces, etc. In this region, interferometric analysis of the IR radiation by means of the FTIR techniques should be preferred.

Additionally, kinetic studies in the time regime of seconds becomes possible due to fast recording and three-dimensional (perspective) display of consecutively recorded spectra [21] and weak or disturbed spectra may be enhanced and corrected. Therefore, FTIR also gains importance in the mIR. Fourier Transform IR Spectrometers which became available several years ago brought real advantages due to the non-dispersive mode of recording the spectra. Since monochromators, slits, etc. are superfluous, the intensity of the exciting radiation can be used in an economical way. Historically, this interferometric method was at the beginning of fIR research [12], but only the recent development of efficient small computers made its broader use possible. In addition to small intensity losses, the rapid recording of FTIR is advantageous for multiple recording and signal-to-noise improvement by averaging.

The core of each (commercial) FTIR spectrometer is a Michelson Interferometer. An interferometer is an apparatus used for splitting and re-unifying light beams so that one of the partial beams can be influenced, e.g. delayed and brought to interference with the second beam. In Michelson's interferometer, the beam is split into two partial beams using a semitransparent mirror (Fig. 7.2). Each of the two beams is reflected at a mirror (one is movable) back to the beam splitter where they are re-unified. The movable mirror creates, either point by point or continuously, all possible phase differences

MICHELSON
INTERFEROMETER

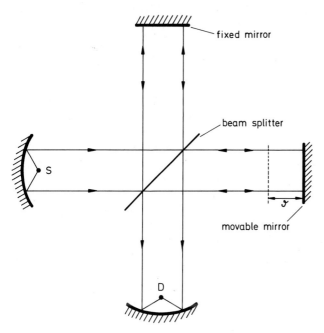

Fig. 7.2. Scheme of a Michelson Interferometer, as used in FTIR spectrometers.
S: Source of IR radiation, *D*: Detector

between the two partial beams. If monochromatic radiation is used, the intensity measured at the detector ($(I(\vartheta)_{\mathrm{rel}})$) is a simple cosine function:

$$I(\vartheta)_{\mathrm{rel}} = B(v') \cos(2\pi v' \vartheta). \tag{7.2}$$

This is due to the fact that the intensities of the two partial beams amplify by interference if

$$\vartheta = 0, \lambda, 2\lambda... \quad \text{and vanish if} \quad \vartheta = \lambda/2, 3\lambda/2.... .$$

Using polychromatic radiation, interference occurs between light of different frequencies so that only at $\vartheta = 0$ we have optimum amplification, whereas at all other differences in path length we observe increasing quenching by interference.

Hence, the resulting interferogram has the form of a damped oscillation. Mathematically, this corresponds to a summation of Eq. (7.2) over all wave numbers:

$$I(\vartheta)_{\mathrm{rel}} = \int_{0}^{\infty} B(v') \cos(2\pi v' \vartheta) dv'. \tag{7.3}$$

$B(v')$, the spectral intensity distribution is the information in which the spectroscopist is interested. It is clearly contained in the interferogram, but in a coded form. Decoding is done mathematically by means of the Fourier transformation, hence the name of the method.

If the Michelson interferometer is the heart of the FTIR spectrometer, the computer is its brain which calculates the spectrum and also performs control functions. The spectra are single-beam spectra, the blank is measured separately and subtracted automatically by the computer. Finally, several runs are averaged and presented as $OD(v')$ or in another familiar format.

7.3.2 IR Reflection and Emission

As a rule, the IR transitions of transparent or soluble polymers are measured using transmission (absorption) techniques, the samples being thin films or solutions in CCl_4, CS_2 or other solvents for limited spectral ranges [13]. In these experiments, reflection is eliminated by conducting the reference beam through a twin cuvette that is filled with the solvent used to dissolve the polymer.

Recording film spectra, reflection cannot be excluded completely, since the reference (if there is any) often has not exactly the same reflection behaviour as the sample. Ideally, a very thin film of the sample polymer should serve as reference, the effective sample thickness being the difference between sample and reference film. Two kinds of polymer samples require special reflection techniques in order to yield useful vibrational spectra:

- Polymer layers on opaque substrates, highly IR scattering samples, certain resins and rubbers not measurable in transmission.
- Oxidised or otherwise surface-modified films, if the surface of the polymer is to be studied in order to elucidate the nature and extent of modification.

Reflection of electromagnetic radiation at smooth surfaces shows maxima in the region of strong absorption bands due to the increase in the refractive index. Diffuse reflection at scattering surfaces, at the other hand, yields maxima in regions of minimal absorption of the sample, since in these spectral regions the losses due to absorption of radiation are very low. The same is true for attenuated total reflection (ATR) of IR radiation, a technique combining total reflection in a crystal, covered by the sample (e.g. a polymer film) and absorption in the thin surface layer which is penetrated by the IR beam (Fig. 7.3). The penetration depth into the medium with lower refractive index (polymer) depends in a complex manner on the wavelength, the refractive indices n_1 and n_2 and on the angle of reflection [15]; as an order of magnitude, the penetration depth can be estimated at $\lambda/10$, i.e. roughly 100 to 1,000 times the value estimated for the surface layer involved in ESCA (Sect. 2.4.2). In practice, multiple reflection is used in order to increase the effective path length (FMIR), Fig. 7.3 b. This method is useful for studying surface modifications of polymer films, e.g. by corona treatment [14] or photochemical oxidation [45]. Experimentally, the greatest difficulty consists in pressing the films (or other samples) homogeneously to the surface of the crystal. Since the optical constants of polymer and crystal vary as a function of the wavelength so that the penetration depth varies, too, the FMIR spectrum is not identical with the absorption spectrum, although the spectral position of the bands is the same.

Finally, IR emission spectroscopy [16] should be mentioned which is based on quantum emission from vibrationally excited states. This emission occurs in competition with non-radiative energy transfer to the surrounding medium and within the sample into other modes of vibration.

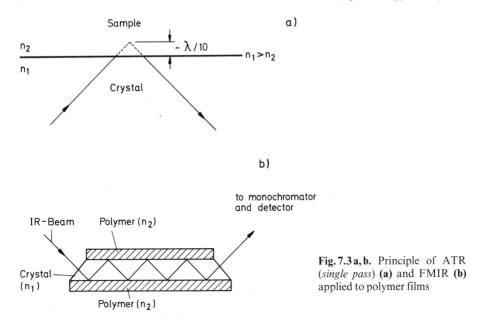

Fig. 7.3a,b. Principle of ATR (*single pass*) **(a)** and FMIR **(b)** applied to polymer films

The average thermal energy at "room-temperature" corresponds to the borderline between mIR and fIR ($kT \approx 1/40$ eV ≈ 200 cm^{-1}). Higher energy vibrations, which are able to exit mIR radiation are excited according to the Boltzmann equation and therefore strongly depend on the average temperature of the sample. Of course, a meaningful analysis of this heat radiation is only possible if the sample does not behave as an ideal black body, a condition which is fulfilled in the case of polymers. This technique could turn out useful for samples whose transmission (absorption) or ATR cannot be measured. The spectral composition of emitted IR radiation is most conveniently measured by FTIR analysis [16].

7.4 Interpretation of Polymer mIR Spectra

7.4.1 Empirical IR Spectroscopy

Empirical IR spectroscopy is based on the existence of "group vibrations" as indicated by narrow frequency ranges of characteristic absorptions. The physical reason for this behaviour is inefficient coupling of the IR absorbing group with the rest of the molecule (Sect. 5.1). This effect is observed in ordinary size molecules as well as in polymers, the former providing an enormous amount of reference spectra [19]. There are, on the other hand, borderline cases to the sphere of "true" polymer spectra caused by vibrations of the whole macromolecule or larger parts of it. In these cases, the vibrating group is influenced by its neighbours so that conformations or crystal structures influence the position, strength or shape of the corresponding absorptions band. These "regularity bands" [22] are usually not observed in amorphous polymers and in solution, except that some ordering exists even in this state, e.g. in helical regions. Accordingly we may distinguish three levels of interpretation, depending on the complexity by the chemical and physical structure of the polymer to be studied (Table 7.1).

C. Vibrational Spectroscopy

Table 7.1. Evaluation of IR spectra

Type of Polymer		Evaluation
A	Complex, irregular chemical and physical structure (e.g. resins, thermosetting resins, random copolymers)	Empirical IR spectroscopy only
B	Intermediate cases (e.g. vinyl polymers with large side groups, homopolymers)	Group frequencies, independent of state of physical ordering "Regularity Bands" indicating regular conformations or even crystallinity
C	Simple chemical structure and thus higher degrees of physical order possible, e.g. helices, extended chains and crystallinity (e.g. PE, POM)	Theoretical band assignments (at least partly) possible

For chemical-analytical purposes, the empirical approach (see also Fig. 5.1) is the most important one [4] and also in polymer physics there is a trend to investigating more complex structures [17] which in part require empirical methods of vibrational analysis.

The assignment of group frequencies is based on reference spectra mostly of monomeric and oligomeric compounds [18, 19]. These can directly be used for interpreting the spectra od amorphous polymers. A comprehensive collection of polymer spectra, including monomers, additives, resins, etc. has been edited by Hummel and Scholl [5]. Several polymer IR and Raman spectra are included in the DMS Atlas [23]. Industrial polymers are delt with by Henniker [57]. These spectra can be used to identify known polymers and plastics by comparison.

The usual nomenclature of vibrations used in empirical IR spectroscopy is explained in Table 7.2. The notation is not always used consistently in the literature. In the following, a few examples of polymer IR spectra will be discussed, knowing that any systematic coverage of the polymers or only groups of polymers studied would surpass the scope of this book.

The field of polymer IR spectroscopy, including applications, has been reviewed several times [4, 5, 17, 21].

7.4.2 Examples of mIR Spectra of Linear Polymers

7.4.2.1 Poly(methylmethacrylate)

PMMA is a glass-forming polymer and the main component of well-known plastics ("PLEXIGLAS") with excellent optical properties. The commercial polymer, which is produced by radicalic polymerisation, was formerly regarded as "atactic" (see Sect.

Table 7.2. Notation of group vibrations

Symbols			Example
A Valence vibrations			
Stretching st		v	
			(sym.; v_s)
			(asym.; v_a)
Breathing br			
B Deformation (Bending) vibrations a)			
Scissor s Vibration		$(\vartheta)^b$	
			(in plane)
Wagging w			
			(out-of-plane)
Rocking r			
			(in plane)
Twisting t			
			(out-of-plane)
Torsion		τ	
Out-of plane		γ	
in-plane-bending		β	

[a] Deformation abbr. *d*, bending abbr. *b*
[b] Often used as general symbol for deformation vibrations

C. Vibrational Spectroscopy

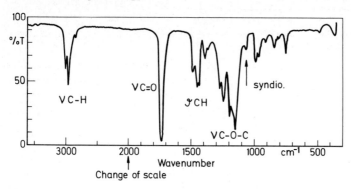

Fig. 7.4. mIR Spectrum of PMMA (radicalic) film after [5]

Table 7.3. Assignment of mIR bands of PMMA (see Fig. 7.4)

Range (see text)	v' (cm^{-1}) (Splitting cm^{-1})	λ (µm)		Assignment
1	3,000	3.33		st CH$_3$
	2,960	3.38		st CH$_3$
	(Maximum)			+ sym. st. CH$_2$
	2,850	3.51		Antisym. st CH$_2$
2	1,735	5.76		st C=O
3	1,490	6.71		
	1,450	6.90		
	1,440	6.94	d C—H	
	1,390	7.19		
	(1,365	7.33)		
4	1,275 ⟩30	7.84		Sym. (?)
	1,245	8.03		st c—o—c
	1,195 ⟩45	8.37		
	1,150	8.70		Antisym. (?)
5	1,060	9.43		Regularity band

10.3). Pure syndiotactic and pure isotactic PMMA can be prepared using ionic initiators.

The mIR spectrum of commercial PMMA is shown in Fig. 7.4 and interpreted in Table 7.3. It consists of four groups of bands:

1) The range of C—H-stretching vibrations is dominated by the Me group $I(CH_3)/I(CH_2) \approx 3$, see formula of basic unit. The central peak (2,960 cm^{-1}) could possibly be resolved using higher dispersion.
2) The C=O stretching range shows a strong homogeneous ester carbonyl band at 1,735 cm^{-1}.
3) CH deformation range showing 4 peaks in the region 6.7 to 7.4 µm.
4) The ester stretching band is a typical "Regularity Band" as can be seen from the behaviour of band splitting [22].

84

Splitting vanishes in copolymers with low MMA-content and decreases with increasing bulkiness of the alkyl group.

Although PMMA is X-ray amorphous, some short-range order seems to be preserved in the amorphous solid, most probably helices. The 1,060 cm^{-1} peak is only found in syndiotactic PMMA indicating a largely syndiotactic configuration of commercial PMMA (see Chap. 10). Configurational regularity (tacticity) facilitates conformational order; the latter can induce crystallinity, but is not bound to do so (as in this case).

7.4.2.2 Poly(isobutene) and amorphous polypropylene

At room temperature PIB is rubber-like, glass-clear and amorphous. It is distinguished from PP by replacing the tertiary H of PP by a methyl group:

$$\left[-CH_2-\overset{\overset{\displaystyle H}{|}}{\underset{\underset{\displaystyle CH_3}{|}}{C^*}}- \right]_n \qquad \left[-CH_2-\overset{\overset{\displaystyle CH_3}{|}}{\underset{\underset{\displaystyle CH_3}{|}}{C}}- \right]_n$$

PP   PIB

The pseudo-asymmetric C* atom of PP is missing in PIB so that no differences due to tacticity are possible in this polymer. The preferred conformation of PIB seems to be a narrow (8/5) helix.

Atactic PP is amorphous and non-oriented, as PIB. It is chemically less stable due to the tertiary CH. Comparing the mIR spectra of the two polymers shows the following (see Fig. 7.5):

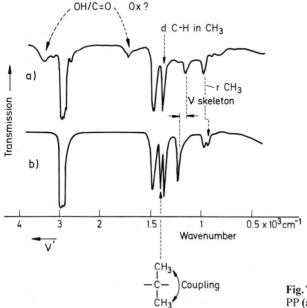

Fig. 7.5 a, b. mIR Spectra of atactic PP **(a)** and amorphous PIB **(b)** films

C. Vibrational Spectroscopy

- Insignificant differences in the C—H valence range (3,000 cm^{-1})
- Weak bands in the OH and C=O valence range for PPat pointing to partial oxidation;
  PIB, on the other hand, is completely free from peaks indicating oxidation products
- In PP the C—H deformation vibration of CH_3 causes a strong single band at 1,378 cm^{-1} (7.26 µm). This peak is split into a doublet in PIB since 2 CH_3 groups are linked to the same C-atom. This effect is called "mechanical coupling" – in contrast to "electrical coupling" which is due to effects of electron distribution (mesomeric effects, etc.). Splitting is quite strong ($\Delta v' \approx 25$ cm^{-1}) despite 2 C—C bonds separate the vibrating C—H groups
- The valence skeleton vibration at 1,153 cm^{-1} (PPat) is shifted to higher frequencies in PIB (1,227 cm^{-1}), whereas
- the r CH_3 (969 cm^{-1} in PPat) is red-shifted in PIB and again split into a doublet. This rocking vibration contains contributions of other modes [24].

7.4.2.3 Poly(vinyl alcohol)

$$\left[-CH_2 - \overset{*}{CH} - \underset{OH}{\big|} - \right]_n$$

PVOH

PVOH is produced by hydrolysis of poly(vinyl acetate) (PVAC) yielding either a completely hydrolysed product (PVOH) or polymers containing residual acetate groups which can be considered as copolymers of vinyl alcohol and vinyl acetate. The mIR spectrum of PVOH is similar to that of its monomeric model substance 2-propanol[24]. Starting from syndiotactic PVAC, partially crystallising stereoregular PVOH can be produced (Fig. 7.6). The broad intense band with maximum of absorption at $v' = 3,370$ cm^{-1} is due to st OH shifted from its position at 3,600 cm^{-1} (2.78 µm, narrow peak) to longer wavelengths and broadened by hydrogen bonding. The formation of hydrogen bonds (O—H...O) evidently weakens the O—H bond; within a series of similar compounds the spectral shift $\Delta v'$ is roughly proportional to the strength of the hydrogen bond[25].

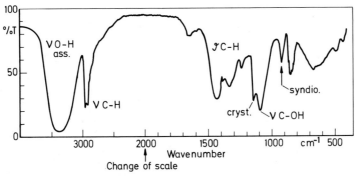

Fig. 7.6. mIR Spectrum of partly crystalline, syndiotactic PVOH, after [5]

The complete absence of any peak at 3,600 cm⁻¹ indicates the absence of any appreciable fraction of free OH groups in solid PVOH. This spectral range can only be evaluated for carefully dried polymer samples (no KBr discs), since water absorbs strongly in the same region and may act both as hydrogen donor and hydrogen acceptor in polymers showing no hydrogen bonds in pure state.

The st CH vibrations in PVOH (Fig 7.6) are found at 2,920 cm⁻¹ (3.42 μm) and 2,950 cm⁻¹ (3.39 μm). The absence of any st C=O absorption near 1,740 cm⁻¹ (5.75 μm) – as found in PVAC [5] – indicates complete hydrolysis of the starting material.

Vibrations of groups containing hydrogen atoms can be by identified replacing H by D. This is especially simple in the case of weakly acidic groups by exchanging their protons by dissolution in D_2O. The frequency of st O—D is reduced with regard to st O—H by $\sqrt{2}$ [Eq. (5.1)]; in our example, this band is expected near 2,400 cm⁻¹ (4.2 μm).

A second hydroxyl band in PVOH is the st C—O of secondary OH groups at 1,090 cm⁻¹ (9.15 μm) near a "crystallinity band" at 1,150 cm⁻¹ (8.7 μm), possibly the st C—O in the crystal phase of the partly crystalline PVOH used for recording Fig. 7.6.

This peak increases by annealing and extensive dehydration over P_2O_5; IR measurements using polarised radiation and oriented samples also indicate the crystalline origin of this band.

An extensive discussion of the PVOH spectrum has been published by Krimm [26] and Tadokoro [27]. According to these authors, the 922 cm⁻¹ (10.85 μm) band is characteristic of syndiotactic sequences.

As in the previous examples, the 1,400 cm⁻¹ range is due to C—H deformation vibrations.

7.4.2.4 Poly(acrylonitrile)

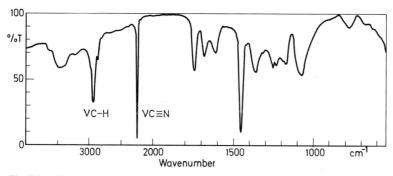

PAN

The most prominent feature of the PAN spectrum (Fig. 7.7) is the st C≡N peak at 2,230 cm⁻¹ (4.48 μm) which in most nitriles appears unvariably at this wave number,

Fig. 7.7. mIR Spectrum of PAN (film)

indicating that this vibration is strongly decoupled from the rest of the molecule. There is furthermore a strong $d\,CH_2$ vibration at 1,447 cm^{-1} (6.91 μm)[28]. The nitrile band is very useful in analysing copolymers (Sect. 7.5.2).

7.4.2.5 Poly(caprolactam)

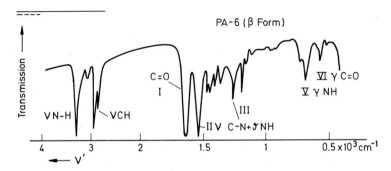

The spectrum of PA-6 serves as an example of the mIR spectra of aliphatic polyamides, which are chemically related to the polypeptides and proteins as far as the amide bond is concerned which links the monomer units. Simple alipathic polyamides are highly crystalline polymers of low molar mass ($M \approx 1$ to 3×10^4); owing to their microcrystallinity they are opaque, tough, and have well defined melting points. As expected from the structural formula, the mIR spectra are composed of bands due to the short $(CH_2)_n$ segments and those which are characteristic of the amide group. This is always planar, since the C(O)N bond is not a pure single bond so that no free rotation is possible. The NH groups from hydrogen bonds with C=O groups of other PA chains. Different crystal structures ("sheet", "pleated sheet", etc.) influence the IR spectrum often to a higher extent than minor differences in chemical structure. The spectrum of the β-form of PA-6 (Fig. 7.8) shows the following features:

The double peak at

| 2,940 cm^{-1} | 3.40 μm | st CH$_2$ antisym. |
| 2,860 cm^{-1} | 3.50 μm | st CH$_2$ sym. |

is typical of all substances whose CH bonds occur predominantly in CH$_2$ groups.

The st N—H at 3,300 cm^{-1} (3.03 μm) is narrow despite its hydrogen bonding (to be compared with PVOH, Fig. 7.6) due to the highly crystalline structure which reduces the fluctuations of length and strength of N—H bonds.

N—H...O=C.

PA-6 (β Form)

Fig. 7.8. mIR Spectrum of PA-6 (β form)

Table 7.4. Characteristic amide bands

No.	Assignment [29]	cm^{-1}	μm
I	st C=O	1,640	6.10
II	st C—N + NH	1,550	6.45
III	st C—N + NH	1,280...50	7.8 to 8.0
IV	d C=O	Not yet detected in PA	
V	γ NH	700	14.3
VI	γ C=O	600	15.6
VII	Lattice vibration	280	36

The amide group causes 7 characteristic vibrations in the range between 1,650 and 250 cm^{-1} summarised in Table 7.4. Six of these have so far been identified in aliphatic PAs. They also appear in the IR and Raman spectra of proteins[29].

Amide (*I*) is shifted to longer wavelengths, since the strength of the C=O bond is weaker than that of a pure double bond.

7.4.3 Tacticity and Conformation in mIR Spectra

As shown in Sect. 7.3, the IR Spectra of complicated polymers are mostly due to group absorption. The interpretation of these spectra is similar to that of low-molar mass compounds. Chemically simple polymers, e.g. PE, PP, POM, show additional bands ("Regularity Bands") which depend on the arrangement of the atoms or groups of atoms in space. The reasons are:
- Formation of well-ordered conformations, including true crystalline order (3D order in regions of at least 5 nm diameter)
- These generally weak "Regularity-Bands" can be observed experimentally since the background of strong, mostly polar group absorptions is weak in simple polymers.

Great influence on the mIR spectra is exerted by conformation, i.e. the 1D order of (linear) polymer chains. High 1D order can, but does not necessarily, lead to true (3D) crystallinity which may influence the IR spectra by factor group splitting. In the example given below (PP), 3D crystallinity seems to have no influence on the spectra which can be explained by "1D" helices and extended chains. In polymers showing pseudo-asymmetric C atoms there is a possible link between configuration (tacticity) and IR spectrum since regular configuration favours or makes possible a conformative order which influences IR absorption.

As an example the IR absorption of PP which can be prepared in isotactic (PP it) and syndiotactic (PP st) form [30-32] will be discussed in the following.

$$
\begin{array}{ccccc}
CH_3 & & CH_3 & & CH_3 \\
| & & | & & | \\
\sim C-CH_2- & C-CH_2- & C-CH_2\sim \\
| & & | & & | \\
H & & H & & H
\end{array}
$$

PP it

In PP it all pseudo-asymmetric C atoms of a chain are either in *d* or *l* configuration.

In the solid state (m.p. 170 °C, density 0.92 g cm^{-3}) the polymer is nearly completely crystalline with a monoclinic elementary cell containing 4 chains or 12 monomeric units

(108 atoms). In organic crystals with large elementary cells, the optical effects depending on unit cell symmetry (Davydov splitting) are weak owing to the large distances of the interacting atoms or groups. More important is the interaction of immediately neighbouring groups, as shown in PP which does not exhibit any Davydov splitting but rather conformational effects due to the 3/1 helix.

In PP st d and l configurations alternate in regular fashion:

$$\begin{array}{ccccc}
CH_3 & & H & & CH_3 \\
| & & | & & | \\
\sim C-CH_2- & C & -CH_2- & C & -CH_2 \sim \\
| & & | & & | \\
H & & CH_3 & & H
\end{array}$$

PP st

This polymer crystallises in two polymorphous forms of slightly different stability. The more stable modification is a double helix where each fourth monomeric unit lies above or below the first. The less stable conformation is the extended chain.

Atactic PP is characterised by an almost statistical pattern of d and l links. It forms a by-product in the industrial production of PP it and, on account its of irregular structure, can neither build helices nor does it crystallise. At room temperature, PP at is sticky and gum-like, it represents a good example of the fact that amorphous solids may form in the absence of stereoregular order. On the other hand, amorphous polymers are not generally and necessarily atactic, as evidenced by PMMA. The spectrum of PP at (Fig. 7.9) is much simpler than those of the more ordered isomers. The differences shown in Fig. 7.9 qualitatively indicate the great influence exerted by coupling between the atoms along the chain. As in the melt, this coupling is only weak in PP at the resulting spectrum consists of typical group absorptions which are identical for the three polymers of equal chemical composition. The IR spectra of PP of different tacticities have been investigated and reviewed by several authors[5, 32–38]. Essential contributions have been made using polarised IR radiation in studying uniaxially drawn PP foils with preferential orientation of the chain axes along the direction of the elongation. Some absorption peaks can only be observed in one direction (parallel or vertical to the chain axis), others show some shifts of the absorption maxima due to orientation-dependent coupling.

The wave numbers of the mIR absorption maxima for PP it and PP st are summarised in Table 7.5. The assignments have been obtained by calculation and comparison with the spectra of partially or fully deuterated PP samples. The following is a brief discussion of the results. For theoretical and experimental details the original papers and extensive reviews should be consulted[33–38].

The C—H valence region (2,800 to 3,000 cm^{-1}) hardly depends on conformation; this indicates that the vibrations are coupled only to a minimum degree (group vibrations).

The 1,430 to 1,460 cm^{-1} region contains one of the strongest bands of the PP spectrum which is caused by the asymmetric deformation vibration of the methyl groups. For the band at 1,460 cm^{-1} (PP it), $\varepsilon_{max} \approx 30$ l mol^{-1} cm^{-1} can be calculated from T and film thickness[5]. The theoretical analysis shows that neither the asymmetric deformation vibration of the CH_3 nor that of the CH_2 group couples with the lattice (backbone) vibrations, thus explaining the similar position and intensity of these bands in PP it, PP st, and PP at. It should be noted that the indices a and b in Table 7.5 refer

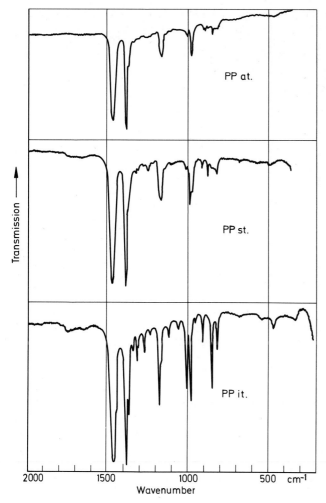

Fig. 7.9. Comparison of mIR Spectra of atactic, syndiotactic and isotactic PP after [5]

to local symmetry coordinates of the helix, which indicates asymmetric vibrations (v and ϑ). The symmetric deformation of the methyl groups – the one causing the strong splitting in PIB (7.4.2.2) – does not couple with the chain and thus again proves to be independent of conformation. The deformation of the tert. H, on the other hand, couples with CH_2 wagging and a lattice vibration; it is pronounced only in the PP it helix (1,350 to 1,380 cm^{-1}).

The next region (1,200 to 1,330 cm^{-1}) contains several weak peaks (hardly observed in thin film spectra) which are distinctly dependent on conformation and thus indirectly on tacticity. The data of this region are those reported by Grant and Ward, quoted after [5]. The 1,150 to 1,170 cm^{-1} region shows the complicated vibrations combining C—C (chain-) valence with different C—H deformation vibrations. Conformational influences, therefore, are not unexpected. The 1,167 cm^{-1} band in PP it is highly charac-

C. Vibrational Spectroscopy

Table 7.5. IR Absorption maxima and assignment of poly(propene)

v'_{max} (cm^{-1})	[a]	PPit	[b]	PPst	[c]
2,958	(s)			2,955	(s)
2,956	(s) ∥ ⊥	v_{as} CH$_3$	(100%)		
2,951	(s) ⊥	v_{as} CH$_3$			
2,923	(s)			2,923	(s)
2,925	(sh) ⊥	v_{as} CH$_2$	(94%)		
2,907	(sh) ⊥	v CH	(94%)		
2,881	(s)				
2,880	(s) ⊥	v_s CH$_3$	(98%)	2,881	(sh)
2,869	(s)			2,869	(sh)
2,868	(s) ⊥	v_s CH$_2$	(99%)		
2,839	(s)			2,841	(s)
2,843	(s) ∥	v_s CH$_2$	(99%)		
2,810	(w)			2,809	(w)
1,458	(s)			1,461	(s)
1,456	(s) ∥	ϑ_b CH$_3$ (65%) $+\vartheta_a$ (CH$_3$) (22%)			
1,450	(s) ∥	ϑ_a CH$_3$ (64%) $-\vartheta_b$ (CH$_3$) (22%)			
1,450	⊥	ϑ_a CH$_3$ (87%)			
1,440	(sh)			1,437	(sh)
1,434	(m) ∥ ⊥	CH$_2$	(91%)		
1,377	(s)			1,376	(s)
1,376	(s) ∥ ⊥	ϑ_s CH$_3$	(77–79%)		
1,360	(m)			1,369	(sh)
1,357	(w) ∥	ϑ CH (50%) $-w$ CH$_2$ (18%)			
1,357	(m) ⊥	ϑ CH (61%), v_a CC (10%)			
1,329	(w)			1,332	(w)
1,326	(vw) ∥	w CH (60%) $-v_b$ CC (21%)			
1,326	(vw) ⊥	w CH (38%), w CH$_2$ (26%) v_b CC (13%)			
1,303	(w)			1,311	(w)
1,305	(w) ∥	w CH$_2$ (57%) $+\vartheta$ CH (29%)			
1,297	(w)			1,290	(w)
1,296	(vw) ⊥	w CH$_2$ (52%), w CH (26%)			
1,255	(w)			1,262	(w)
1,255	(w) ∥	t CH$_2$ (80%)			
1,219	(w)			1,242	(w)
1,218	(vw) ⊥	t CH$_2$ (75%), w CH (12%)			
				1,200	(w)
1,168	(s)			1,162	(m)
1,166	(m) ∥	v_b CC (26%) $-r_a$CH$_3$ (18%) $+$ w CH (18%)			
1,153	(sh)			1,154	(m)
1,153	(sh) ⊥	v CC (17%), v_b CC (13%), ϑ CH (8%)			
1,103	(w)			1,128	
				1,083	(w) to

Table 7.5 (continued)

v'_{max} (cm^{-1})	[a]	PPit	[b]	PPst	[c]
1,101	(vw) $\perp$	v_b CC (25%), r_a CH$_3$ 16% r_b CH$_3$ (11%)		1,061	(vw)
1,045	(w)				
1,043	(vw) $\parallel$	v_a CC (53%)			
997	(s)				
998	(m) $\parallel$	r_b CH$_3$ (41%)		1,002	(w) to (vw)
972	(s)			976	(s)
974	(m) $\parallel$	r_a CH$_3$ (55%) $+ v_b$ CC (26%)		963	(m)

[a] PPit intensity and polarisation relative to drawing axis;
[b] % calculated contribution in coupled vibration
[c] PPst, intensity (helix)

teristic of the 3/1 helix; it is absent in the spectra of the melt and of PP st and thus represents a typical and specific chain (lattice) vibration.

There are weak lattice vibrations which are drastically different in PP it and PP st, in the region from 1,000 to 1,130 cm^{-1}. Again, from 950 to 1,000 cm^{-1}, a characteristic coupling between CH$_3$ rocking and C—C valence vibrations depends on conformation and yields strong differences between PP it and PP st. The strong PP it peak at 997 cm^{-1} (3/1 helix) is missing in the spectrum of PP st. The CH$_3$-rocking band at 973 cm^{-1}, on the other hand, occurs in both isomers and can conveniently be used as internal standard in studying the degree of crystallinity[5]. The last lowest wave number absorption peaks which are reasonably strong are in the range between 800 and 950 cm^{-1}. Important features are the medium intensity band at 868 cm^{-1} in PP st, characteristic of the double helix in this isomer and the nearly pure CH rocking at 841 cm^{-1}, characteristic of the 3/1 helix of PP it. Below this range there are several very weak conformation-dependent peaks down to 200 cm^{-1} [5].

Finally, it should be pointed out that a (nearly) complete assignment of conformation-dependent IR spectra requires carefully prepared and purified polymer samples of well-defined tacticity, helical order and – for the sake of comparison – deuterated polymers. Since, as we have seen, the conformation-dependent absorption peaks are often quite weak, impurities may be especially misleading in these studies.

Calculations of IR spectra of crystalline, linear polymers have shortly been discussed in Chap. 5. The basis of these calculation is the normal coordinate analysis[56]. If a family of polymers shows similar conformation in the crystalline phase, a common valence force field may be used, introducing a few additional parameters for each specific member of the family.

This has been demonstrated for (trans)1,4-poly(butadiene)[58], (trans)1,4-poly(2,3-dimethylbutadiene), and (trans)1,4-poly(isoprene)[59]. The combined information of IR, Raman and theory has been used in order to elucidate the preferred conformation of crystalline PVCl$_2$[60].

7.5 Applications of mIR Spectroscopy

7.5.1 General

The possibilities for applying mIR spectroscopy in polymer research are so numerous that a systematic treatment is not feasible here. In the following, only a few important applications will be mentioned. More extensive discussions of practical applications are given in the often quoted books by Hummel[4], Hummel-Scholl (first edition)[5], the chapters devoted to analytical applications in Zbinden[39] and in the most recent book by Siesler and Holland-Moritz[21]; see also the comprehensive recent review by Snyder[17]. Special attention to FTIR in polymer research is paid in [21] and [40]. it is a distinct advantage of this technique over conventional dispersive IR spectroscopy that moderately rapid reactions can be followed by recording a spectrum every few seconds if highest resolution is not required. The presentation can be made very pictorial by "3D" (perspectivic) plotting, showing immediately the growing in or vanishing of certain groups as a function of time[21, 40].

7.5.2 Copolymers

The composition of copolymers using mIR spectra can be determined if at least one partner (two in the case of terpolymers) shows a characteristic and reasonably strong absorption band in an otherwise free spectral region. If the molar ratio A/B of the monomers A and B is small, A has to be the IR-detectable component. A real "text-

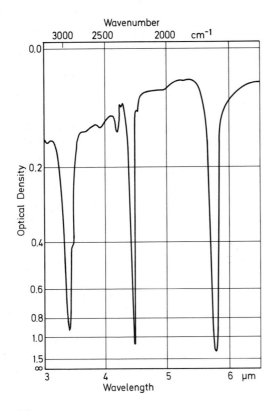

Fig. 7.10. Infrared spectrum of poly (acrylonitrile-co-methylmethacrylate) (94/6) after[39]

book example" is offered by the copolymer derived from methyl methacrylate ($v > C = O$ band) and acrylonitrile ($v -C \equiv N$ band), see Fig. 7.10. Please note the strongly different molar absorption coefficients of the two key peaks, the nitrile triplebond absorption being roughly 10 times weaker than the corresponding carbonyl double bond absorption. The range of A/B which can be studied by IR spectroscopy amounts to about 0.01 to 100 in favourable cases, in others roughly 0.1 to 10. Knowing the film thickness (or polymer concentration in case of solutions) the evaluation can be performed using Lambert-Beer's law. Alternatively, relative optical densities can be used by selecting a strong band of A or B as internal standard.

If sequences of the AAA ... or BBB ... type cause modifications of the spectrum, e.g. by conformational order, block-copolymers can in principle be distinguished from random ones.

7.5.3 Molar Mass

If one or both end groups of a linear polymer are known, e.g. if the initiator and the polymerisation mechanism are known, the number average $\bar{M}_n$ of the relative molar mass can in principle be measured by IR. The number of end groups (N_E) corresponds to the number of macromolecules N_P, since $N_P = N_E$ or $N_P = N_E/2$, thus yielding the information necessary for calculating $\bar{M}_n$. The method of determining N_E and thus $\bar{M}_n$ by IR absorption requires a distinct absorption band of the end group(s) and is limited to oligomers up to about $\bar{M}_n = 10^4$.

7.5.4 Branching

Polymers showing irregular side chains (e.g. LDPE) are not suitable for determining $\bar{M}_n$ by IR absorption. The degree of branching, however, can be measured if the end groups of the side chains (E_3) differ significantly from those of the main chain (E_1, E_2) either with regard to their chemical composition or due to much higher concentration $[E_3] \gg [E_1] = [E_2]$; see Fig. 7.11. Alternatively, changes in the main chain (branching points as indicated schematically in Fig. 7.11) can be used in order to estimate the degree of branching. Branching requires, e.g., the replacement of a CH_2 group by a CH group. Of course, any other sensitive method which distinguishes the branching point from the unsubstituted chain segment can be used as well, e.g. ^{13}C-NMR.

As in the above methods, the concentration (end groups or branching points) can be determined by using Lambert Beer's Law or an internal reference peak.

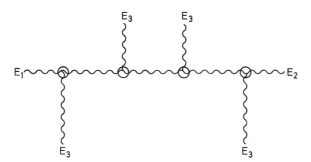

Fig. 7.11. Scheme of a branched linear polymer

Table 7.6. Wave number and intensity of absorption bands characteristics of various kinds of double bonds after Zbinden [39]

	C=C (Stretching)		C—H (Out-of-Plane bending)	
	(cm^{-1})	ε [a]	(cm^{-1})	ε [a]
R—CH=CH$_2$, two C—H	$1{,}642 \pm 3$	28–44	990–1,003	33– 57
bands			908– 916	110–150
1,2-polyisoprene			909	149
1,2-polybutadiene			911	145
R—CH=CH—R′ (trans)	1,667	Very weak	964– 977	85–120
trans-1,4-polybutadiene			967	109
R—CH=CH—R′ (cis)	$1{,}646 \pm 10$	7–18	675– 729	13–106
cis-1,4-polybutadiene			680	23
RR′C=CH$_2$	$1{,}650 \pm 11$	20–42	883– 895	103–200
3,4-polyisoprene			888	145–159
RR′C=CHR″	$1{,}680 \pm 10$	4– 7	788– 840	11– 35
cis-1,4-polyisoprene			835	19
trans-1,4-polyisoprene			842	11

[a] $1\ mol(\text{double bond})^{-1}\ cm^{-1}$

7.5.5 Carbon Double Bonds

Unsaturated groups belong to those impurities or end groups which are most frequently encountered in formally saturated polymers. In polymers derived from monomers having conjugated double bonds, C=C groups may occur in the repetitive units of the polymer, either in the main chain (1,4 addition) or in the side groups (1,2 addition). Characteristic absorption peaks can be found near 1,640 to 1,680 cm^{-1} (νC=C) and 680 to 1,000 cm^{-1} (γC—H), as measured in isomeric poly(butadienes) and poly(isoprenes) [39]. Natural caoutchouc consists mainly of cis-poly(isoprene).

γC—H: 835 cm^{-1}; ε: 19 l mol (double bond)$^{-1}$ cm^{-1}.

Further data are compiled in Table 7.6. Double bonds in PVC, together with syndiotactic sequences can be measured in agreement with other methods [61].

7.5.6 Oxidation Processes

Chemical modifications of polymers are very frequently studied by means of mIR absorption and ATR (FMIR) spectroscopy. These modifications can be applied purposely in order to ameliorate a polymer or they may occur during production and use under the influence of UV-radiation, elevated temperature, etc. The most important degradation pathways are oxidative processes caused by

– UV-radiation + oxygen ($^3O_2$)
– Heat + oxygen ($^3O_2$)
– Ozone (O_3)
– Molecular singlet oxygen ($^1O_2$)

In most cases, strongly polar groups are formed which can easily be detected by mIR spectroscopy.

Hydroperoxy groups (R—OOH) are the most important primary products of thermooxidative and often also of photo-oxidative degradation processes[41]. These groups are unstable, both thermally and photochemically and are converted to alkoxy radicals which react further to give hydroxyl and carbonyl groups, both absorbing strongly in the mIR.

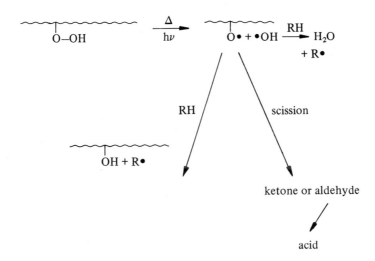

Unfortunately, vOO—H can hardly be distinguished from vO—H in the mIR. Marked differences can be observed between free and associated (H-bonded) OOH groups. At this point it is interesting to note that PP invariably shows associated OOH groups even at very low levels of oxidation[42]. In contrast to this strange behaviour, PE shows free —OOH groups at low concentration in the 2.5 to 3.0 μm region. Chien et al.[42] concluded that —OOH groups in PP are present in 1,3- or 1,3,5- etc. sequences forming intramolecular hydrogen bonds, as postulated earlier from kinetic measurements in solution[43]. There is no direct evidence of structures of this type in solid PP, although the mIR spectra strongly suggest a similar behaviour in both phases.

Provided there are no carbonyl or hydroxyl groups in the basic unit of the polymer, oxidation can be detected by monitoring the changes in the O—H and C=O regions (Fig. 7.12).

In many cases, oxidative degradation starts at the surface[44, 45], e.g. during extrusion of films at high temperature and in contact with air. In this case, ketones are formed at the surface which photochemically react by Norrish I and Norrish II splitting, thus decreasing $\bar{M}$ and consequently the mechanical strength. Finally, also the optical properties of the product are affected. Studying these processes, the mIR-ATR (-FMIR) spectra should be recorded in addition to mIR transmission spectra[45] in or-

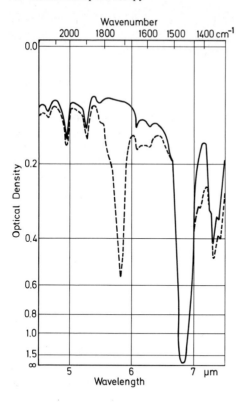

Fig. 7.12. Infrared spectra of films of PE; exposed to UV radiation in the presence of oxygen and water vapour (– – –); unexposed film (——) after [39]

der to investigate specifically the surface layer (see 7.3.2); e.g., the surface layers of PET have been studied during photochemical degradation [62]. In this case, fluorescent groups are formed at the surface which may also detected by luminescence spectroscopy.

7.5.7 Corona and Plasma Treatment of Polymer Surfaces

A method of improving certain properties such as adhesion, printability, wettability, etc. is the deliberate, controlled degradation and cross-linking of polymer surfaces. This can be done by purely chemical methods (e.g. ozone, sodium/NH_3 depending on the chemical nature of the polymer) or by corona discharge, a special type of plasma which is generated by accelerated electrons leaving a wire or an edge under high tension. The active particles formed – atoms, ions, radicals, O_3 and NO_x – modify the surface chemically. Again, ATR is an adequate method for studying the changes [45] in surface layers of about 1 μm (much thinner layers of the order of 1 nm can be studied by ESCA, see Sect. 2.4.4).

Using the ATR technique, carbon double bonds have been identified at 1,640 cm^{-1} under nitrogen gas [14]. In an oxygen atmosphere carbonyl compounds are formed in addition, probably by ozone formed in a secondary reaction. Unfortunately, the ATR technique is not sensitive enough in order to detect submonolayer modifications known to modify the adhesive properties of the polymer surfaces. In these cases, either ESCA or specific colour-forming reactions can be used, e.g. 2,4-dinitrophenyl-hydrazine for ketones and aldehydes [63], and measured by UV-absorption (Chap. 3). Chemical treatment by halogens etc. has also been studied by means of IR/ATR [64].

7.5.8 Dissociation of Polymeric Acids and Other Polymer-IR Studies

Polystyrene sulphonic acid, a strong acid acting as cation exchanger, can be investigated in the mIR in the form of thin films. Dissociation starts if two mol H_2O have been taken up per mol SO_3H [46]. In the IR spectrum the absorption of the $H_5O_2^+$ ($H_3O^+ + H_2O$) unit formed can be detected as an extremely broad band which is explained by rapid tunneling of the proton in the highly polarisable H bond connecting the two subunits. This behaviour is to be distinguished from simple association by H bonds which leads to red-shifted absorption, e.g. in PVOH (Sect. 7.4.2.3). Other recent studies, mostly using FTIR include time-dependent orientation processes [65, 66], polymer blends [67, 68] and the formation of carbon fibers by thermal degradation of linear polymers, e.g. PAN [69].

7.6 Résumé of mIR

The main application of mIR polymer spectroscopy is doubtless the analysis and identification of chemical structures and their changes (Table 7.7). This application is followed, at a higher level of sophistication, by studies of conformation which also provide information about tacticity and crystallinity. True crystal effects caused by 3D ordering are occasionally also observed, although the effects are small. The electronic structure influences the strength of many bonds and thus causes shifts of maxima which may help to elucidate the nature of bonding.

Many mIR bands are due to movements in the chain and can thus be interpreted as phonons (see Sect. 5.6).

As a prominent example for interactions and complex formation studied by mIR spectroscopy, the H bond has been discussed on several occasions. Summing up, mIR spectroscopy is the most commonly used method in polymer spectroscopy. This is why only a very limited number of examples out of thousands has been discussed in this chapter. This extensive use as a routine method overshadows the scientific merits of the method which are often underestimated [56]. The range of applications has recently been increased by FTIR.

Table 7.7. Information obtained by mIR absorption relating to structure and dynamics of polymeric systems

Structure		Dynamics	
Chemical structure	+	Movements of the chain,	+
Tacticity	(+)	segments and side groups	
Conformation	+	Phonons	+
Crystallinity	+	Excitons	−
Electronic structure	+	Complex formation and related phenomena	+

7.7 Far-Infrared Spectroscopy of Polymers

7.7.1 Introduction

The far infrared extends from $v' = 200$ to 10 cm^{-1} (IUPAC) or $\lambda = 50$ to $1{,}000$ μm. Since there is no unambiguous physically meaningful borderline between medium and far IR[1], in older papers the fIR is said to extend up to 300 or even 500 cm^{-1}.

Absorption by polymers in the fIR[47] is due to low-frequency vibrations (rotations cannot be observed in polymers) of heavy groups and/or weak bonds [Eq. (5.1)]:

- An important vibration in the fIR consists of the valence vibration of H-bonds X— H...Y, not to be confused with vX—H and ϑX—H, modified by H-bonds and observed in the mIR
- Intermolecular vibrations in DA complexes
- Torsional vibrations
- C-Halogen vibrations
- Phonons (quanta of collective lattice vibrations).

The recording of fIR spectra has greatly been facilitated by the FTIR technique (Sect. 7.3.1), using evacuated interferometer and sample housing in order to avoid strong absorption caused by water vapour.

The absorption coefficients of fIR transitions in polymers are often very weak (actually LDPE is a popular material for windows and cuvettes, PET films are used as beam splitter in FTIR spectrometers).

Competing with fIR-absorption spectroscopy is Raman spectroscopy (region near v'_0) and inelastic neutron scattering. The latter method, using particle radiation rather than electromagnetic waves, has other selection rules, since momentum transfer favours collision partners of roughly the same mass.

7.7.2 Chemical Applications

The low-frequency absorption of the H bond in the fIR is due to vibrations of the molecules or connected by the H bond. The frequency of this band is therefore only weakly influenced by deuteration in contrast to mIR bands caused by X—H/X—D vibrations.

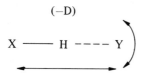

An example of fIR absorption due to an H bond is shown in Fig. 7.13. Actually, the polymer (PE) serves as a solvent in this experiment which demonstrates (a) the reversible association of 2,5-dimethylphenol and (b) the unambiguous identification of the 170 cm^{-1} band to belong to the intermolecular H bond[48].

---

1 Incidentally, the borderline defined by IUPAC corresponds, in energy units, to the average thermal energy at room temperature,

$$50 \text{ μm} \cong 200 \text{ cm}^{-1} \cong 1/40 \text{ eV} \approx kT.$$

In the fIR, and even more so in spin resonance spectroscopy (Part D), the excited states are highly populated at ordinary temperature

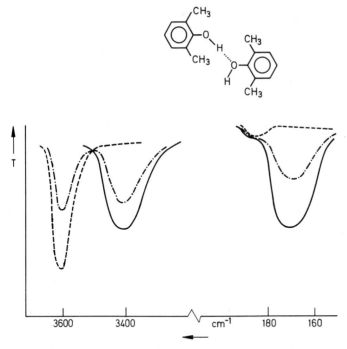

Fig. 7.13. fIR Spectra of 2,6-dimethyl phenol dissolved in PE at room temperature (——), after heating to 100 °C and rapid cooling (– – –) and after some time at room temperature (–·–·–) after [48]

One of the more complex polymers investigated by fIR, PET [49], shows four to six peaks in the region 40 to 200 cm^{-1} if uniaxially stretched (100 μm film) and measured using polarised IR.

There are considerable differences between amorphous and crystalline samples, indicating strong intermolecular interactions.

Biaxially stretched crystalline PET films show three peaks at 68, 78, and 176 cm^{-1} which are absent in amorphous samples. They have been interpreted as being due to phonons.

Poly(vinylidene fluoride) (PVF$_2$) forms three different solid phases, depending on the mode of preparation, one of the phases (I or β) being piezoelectric. The polarised fIR spectra of the three phases have been measured and compared with mIR and X-ray results [50]. The spectra are very sensitive to the arrangement of the polymer chains and to deviations from the ideal crystalline order. They can be used to identify the different phases and distinguish between samples of high and low degree of purity. Furthermore, it is hoped to understand the transition between the phases by extended studies of fIR spectra. The energy of fIR vibrations being small, weak interactions typical of intermolecular forces can easily be recognised.

7.7.3 Phonons in fIR Spectra of Polymers

Phonons have been discussed in Sect. 5.6. In the spectral region of fIR, the low-energy intermolecular vibrations of crystalline regions can be described as (optical) phonon

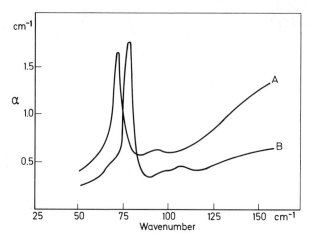

Fig. 7.14 A, B. fIR Spectrum of HDPE (90% crystalline) at 25 °C **(A)** and −150 °C **(B)**, after [52]

transitions. Thus, part of the absorption peaks mentioned in Sect. 7.7.2 (PET, PVF$_2$) are due to phonons. The selection rule for optical absorption ($\phi \approx 0$) means that the wavelength of the phonons – in the wave picture phonons are considered as collective lattice vibrations – is much larger than the lattice constant.

An example of polymer phonon absorption in the fIR is illustrated in Fig. 7.14 (HDPE) [51, 52]. The absorption is shown at two temperatures. There is a sharp peak at 72 cm^{-1} at 123 K. The lowest frequency internal (intramolecular) vibration of PE is calculated to have ten times the observed frequency. Hence, the absorption observed has to be due to an intermolecular vibration. Its intensity is furthermore decreased considerably in LDPE of medium crystallinity, indicating the origin of the 72 cm^{-1} peak in the crystal regions of PE. Cooling of the sample increases the density of the polymer and causes the intermolecular (repulsive) potentials to become steeper; this explains the increased wave number at low temperature at which there are fewer vibrationally excited molecules so that transitions from vibrating polymers to higher vibrational states do not occur. Thus, the spectrum becomes sharper.

Poly(tetrafluoroethylene) (PTFE) is another example of phonon absorption [51, 53]. This polymer shows only diffuse absorption at room temperature. At low temperature, however, crystalline PTFE shows a well-resolved spectrum consisting of at least four peaks. There is a phase transition in PTFE near room temperature so that the low temperature spectrum not only shows the usual band sharpening due to the loss of excited state absorptions and steepening of potentials, but also reflects a different crystal structure. There is an indication of splitting of the 55 cm^{-1} peak, which is said to be due to the presence of two PTFE chains per unit cell [51]. Other modes in the fIR spectrum correspond to calculated internal modes.

7.7.4 Inelastic Neutron Scattering (INS)

"Inelastic neutron scattering is used to obtain molecular spectra of polymers free from the restrictions of optical selection rule" [54]. The usefulness of INS justifies the discussion of this particle radiation which is partly out of the scope of this book devoted to the interaction of electromagnetic radiation with polymers (Chap. 1).

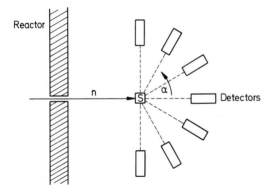

Reactor

n

α

Detectors

Fig. 7.15. Scheme of a neutron scattering experiment: *n*: neutron beam, *s*: sample, α: scattering angle

The neutron has the following properties:
- as particle
  - Mass: 1 atomic unit $= 1.675 \times 10^{-27}$ kg
  - Electric charge: 0
  - Spin: $1/2\,(h/2\pi)$
- as wave (at a velocity of $v \approx 800$ ms^{-1})
  - de Broglie wavelength: 0.5 nm.

At the given velocity, neutrons have a kinetic energy of 300 J mol^{-1} (corresponding to a 23 cm^{-1} photon) and a momentum of 1.3×10^{-24} Ns (an energy equivalent photon has a momentum of only 1.5×10^{-30} Ns). Neutrons have a much higher momentum and a much shorter wavelength compared to photons of the same energy per particle.

These differences lead to the following characteristics of INS:
- Scattering involves the nuclei rather than the electron
- The cross-section strongly depends on the isotopes (drastic differences between H and D)
- Scattering can occur in a spin-coherent and a spin-incoherent manner depending on the spin properties of the nuclei.

The unique position of INS in phonon research is due to the fact that energy, momentum and wavelength are in a suitable range for measuring phonons.

Information on structure and dynamics can therefore be obtained at one and the same radiation. The transfer of energy and momentum occurs in an "anti-Stokes" manner from polymer (i.e. from the phonon) to the neutron whose kinetic energy increases and whose de Broglie wavelength consequently decreases. Both effects can be used to measure the amount of quantum energy transferred. The scattered radiation is measured as a function of the angle relative to the direction of the exciting beam (Fig. 7.15).

If the velocity of the neutrons is measured ($E_{kin} = mv^2/2$), a time-of-flight spectrometer is used as detector. Changes in wavelength are measured by crystal diffraction, using Bragg's law ($\lambda = 2a \sin \phi$ where $a =$ lattice constant, $\phi =$ angle of maximum reflection). In the extreme case of coherent INS, one experiment combines inelastic scattering and diffraction (interference) so that a high density of information is obtained. For this experiment, perdeuterated polymers are required, whereas for incoherent INS protonated polymers are more suitable. This is due to the different cross-sections of scattering (Table 7.8).

C. Vibrational Spectroscopy

Table 7.8. Cross-sections for neutron scattering (low energy neutrons) [54]

Nucleus	Spin $(h/2\pi)$	Coherent (barn)[a]	Incoherent (barn)[a]
^{1}H	1/2	1.8	80
^{2}H	1	5.6	2
^{12}C	0	5.6	0
^{14}N	1	11.6	0.3
^{16}O	0	4.2	0
^{28}Si	0	2.0	0

[a] 1 barn $= 10^{-24}$ cm^2/nucleus

Coherent INS requires large single crystals of deuterated polymers which are available only for a few polymers (e.g. POM [55]). The information to be gained, especially for transversal phonons, cannot be obtained by other methods, i.e. the intermolecular part of the crystal potential. This is needed for calculations of polymer conformations and elastic behaviour of crystalline polymers.

The experiments are simpler if oriented partly crystalline polymers (e.g. fibres) are used instead of single crystals, PTFE being a good example for this type of experiment [55]. PTFE crystallines in a hexagonal lattice, the macromolecules forming nearly ideal cylinders (15/7 helix at room temperature), see Fig. 7.16. The helices are arranged in the direction of the fibres.

The time-of-flight spectra show three maxima; two maxima at $1/v = 400$ and $1/v = 1,200$ µs m^{-1} are independent of the scattering angle, the latter corresponding to elastically scattered neutrons ($v = 800$ ms^{-1}, $\lambda = 0.5$ nm). The third peak (about 800 µs m^{-1} at 27°) approaches the elastic peak near Bragg's angle, i.e. The energy transfer from the phonons of PTFE ceases if the diffraction relation is fulfilled. This experiment shows that phonon dispersion (see Fig. 5.3 in Sect. 5.6) is actually measured which is not possible in optical experiments.

The reason for the dependence of scattering intensity on the angle of scattering is the angle dependence of momentum (mv) transfer which is related with the Broglie's wavelength (λ)

$$mv = h/\lambda. \tag{7.4}$$

Since energy transfer from the phonons of the polymer in this special case (Fig. 7.16) discontinues at Bragg's angle ($\phi = 0$), the phonons have zero frequency. This indicates

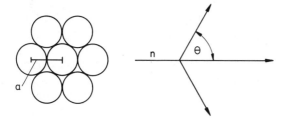

Fig. 7.16. Coherent INS by hexagonal PTFE fibers: a: lateral distance of PTFE helices, θ: Bragg's angle (56.5°), n: neutron beam

Table 7.9. Information obtained by fIR and INS relating to structure and dynamics of polymeric systems

Structure		Dynamics	
Chemical structure	(+)	Movements of the chain,	+
Tacticity	(+)	segments and side groups	
Conformation	+	Phonons	+
Crystallinity	+	Excitons	−
Electronic structure	−	Complex formation and re-	+
		lated phenomena	

acoustic phonons running vertical to the helix axis of the oriented PTFE molecules. In contrast, the fIR experiment (7.3.3) populated the optical branch (at $\Delta\phi = 0$) of the dispersion curve.

7.7.5 Résumé of fIR + INS

The most interesting results with fIR and INS spectroscopy can be obtained on phonons, chain movements and intermolecular interactions (Table 7.9). These processes critically depend on crystallinity and conformation. Chemical-analytical applications are much less important compared to mIR. It should be noted, however, that modern FTIR spectrometers have become available only recently and that the expansive instrumentation of INS, requiring a nuclear reactor as radiation source, forbid any routine application.

References

1. Coblentz, W.W.: Investigations of Infrared Spectra, Publications Nos. 35, 65, 97, Washington: Carnegie Institution 1905–1908
2. Lecomte, J.: Spectroscopie dans l'infrarouge, in: Handbuch der Physik XXVI, Berlin, Göttingen, Heidelberg: Springer 1958, p. 244
3. Brügel, W.: Einführung in die Ultrarotspektroskopie, 4.Ed. Darmstadt: Steinkopff 1969; Polymers: p. 356
4. Hummel, D.O., in: Hummel, D.O. (ed.): Polymer Spectroscopy, Weinheim, Verlag Chemie 1974, p. 112
5. Hummel, DO., Scholl, F.: Atlas der Polymer- und Kunststoffanalyse, 2nd ed. München: Hanser, Weinheim: Verlag Chemie (since 1978); 1st ed. (1968–1973)
6. Wheeler, O.W.: Chem. Rev. 59, 635 (1959)
7. Brown, T.L.: Chem. Rev. 58, 581 (1958); J. Am. Chem. Soc. 80, 3513 (1958)
8. Ref. [5], 1st ed., Vol. 1, Part 1, p. 48 (1968)
9. Goddu, R.F., Delker, D.A.: Anal. Chem. 32, 140 (1960)
10. Ferraro, J.R., Basile, L.J. (eds.): Fourier Transform Infrared Spectroscopy, Vol. 1, New York: Academic Press (1978), Vol. 2 (1979), Vol. 3 (1982)
11. Barnes, A.J., in: Vibrational Spectroscopy, Modern Trends, Barnes, A.J., Orville-Thomas, W.J. (eds.), Amsterdam: Elsevier 1977, p. 53
12. Durig, J.R., Cox, A.W., Jr., in (10), Vol. 1, p. 215
13. White, R.G.: Handbook of Industrial Infrared Analysis, Plenum, New York (1964); Tables of useful solvents for IR spectroscopy, p. 102
14. Carlsson, D.J., Wiles, D.M.: Can. J. Chem. 48, 2397 (1979)
15. Harrick, N.J., du Pré, F.K.: Appl. Optics 5, 1739 (1966)

C. Vibrational Spectroscopy

16. Bates, J.B.: Infrared Emission Spectroscopy, in: Fourier Transform Infrared Spectroscopy, Vol. 1, p. 99 ff., Ferraro, J.R., Basile, L.J. (eds.), New York: Academic Press 1978
17. Snyder, R.-G., in: Fava, R.A. (ed.): Methods of Experimental Physics, Vol. 16, Polymers, Part A, Molecular Structure and Dynamics, New York: Academic Press 1980
18. Bellamy, L.J.: Ultrarotspektrum und chemische Konstitution, 2nd ed., Darmstadt: Steinkopff 1974, Engl. Original ed.: The Infra-red Spectra of Complex Molecules, 2nd ed., London: Methuen
19. Sadtler: Standard Infrared Spectra, Philadelphia
20. CRC Handbook of Chemistry and Physics, 58th ed. 1977–1978, Section F, Collection of Identification Tables
21. Siesler, H.W., Holland-Moritz, K.: Infrared and Raman Spectroscopy of Polymers, New York: Marcel Dekker 1980
22. Zerbi, G., Ciampelli, F., Zamboni, V.: J. Pol. Sci. C-7, 141 (1964)
23. Schrader, B., Meier, W. (eds.):DMS Raman/IR Atlas of Organic Compounds, Weinheim: Verlag Chemie, Vol. 2, Section N 1974–1975
24. Ref. [5], 1st ed., Vol. I/1 (1968)
25. Hadži, D. (ed.): Hydrogen Bonding, New York: Pergamon 1959
26. Krimm, S., Liang, C.Y., Sutherland, G.B.B.M.: J. Pol. Sci. 22, 227 (1956)
27. Tadokoro, H., et al.: J. Chem. Phys. 23, 1351 (1955) and J. Pol. Sci. 26, 379 (1957)
28. Yamahashi, R., Tadokoro, H., Marahashi, S.: J. Chem. Phys. 41, 1233 (1964)
29. Keighley, J.H., in: Jones, D.W. (ed.): Introduction to the Spectroscopy of Biological Polymers, London: Academic Press 1976, p. 17
30. Natta, G.: Makromol-Chem. 35, 44 (1960)
31. Natta, G., Peraldo, M., Allagra, G.: Makromol. Chem. 75, 215 (1964)
32. Miyazawa, T., in: The Stereochemistry of Macromolecules, Vol. 3, Kethley, A.D. (ed.), New York: Dekker 1968, p. 147
33. Tadokoro, H., Kobayashi, M., in: Hummel, D.O. (ed.): Polymer Spectroscopy, Weinheim: Verlag Chemie 1974, p. 3
34. Natta, G., Valvassori, A., Ciampelli, F., Mazzanti, G.: J. Pol. Sci. A3, 1 (1965)
35. Schachtschneider, J.H., Snyder, R.G.: Spectrochim. Acta 21, 1527 (1965)
36. Tadokoro, H., Kobayashi, M., Uhita, M., Yasufuku, K., Murahashi, S.: J. Chem. Phys. 42, 1422 (1965)
37. Peraldo, M., Cambini, M.: Spectrochim. Acta 21, 1509 (1965)
38. Grant, I.J., Ward, I.M.: Polymer 6, 223 (1965)
39. Zbinden, R.: Infrared Spectroscopy of High Polymers, New York: Academic Press 1964
40. Siesler, H.W.: J. Mol. Structure 59, 15 (1980)
41. Geuskens, G. (ed.): Degradation and Stabilisation of Polymers, London: Applied Science Publ. 1975
42. Chien, J.C.W., Vandenberg, E.J., Jabloner, H.: J. Pol. Sci. A-1 6, 381 (1968)
43. Dulog, L., Radlmann; E., Kern, W.: Makromol. Chem 60, 1 (1963)
44. Carlsson, D.J., Wiles, D.M.: Macromolecules 4, 174 (1971)
45. Blais, P., Carlsson, P.J., Wiles, D.M.: J. Pol. Sci. A-1, 1077 (1972)
46. Zundel, G.: Hydration and Intermolecular Interaction, Infrared Investigations with Polyelectrolyte Membranes, New York: Academic Press 1969
47. Brasch, J.W., Mikawa, Y., Jakobsen, R.J.: Appl. Spectrosc. Rev. 1, 187 (1968)
48. Jakobsen, R.J., Brasch, J.W.: Spectrochim. Acta 21, 1753 (1965)
49. Maubey, T.R., Williams, D.A.: J. Pol. Sci. C22, 1009 (1969)
50. Latour, M., Montaner, A., Galtier, M., Geneves, G.: J. Pol. Sci. Pol. Phys. Ed. 19, 1121 (1981)
51. Willis, H.A., Andby, M.E.A., in: Ivin, K.J. (ed.): Structural Studies of Macromolecules by Spectroscopic Methods, London: Wiley 1976, p. 81
52. Fleming, J.W., Chantry, G.W., Turner, P.A., Nicol, E.A., Willis, H.A., Cudby, M.E.A.: Chem. Phys. Lett. 17, 84 (1972)
53. Chantry, G.W., Fleming, J.W., Nicol, E.A., Willis, H.A., Cudby, M.E.A., Boerio, F.J.: Polymer 15, 69 (1974)
54. Allen, G., in: Ivin, K.J. (ed.): Structural Studies of Macromolecules by Spectroscopy Methods, London: Wiley 1976, p. 1

55. White, J.W.: in ref. [54)], p. 41
56. Painter, G.C., Coleman, M.M., Koenig, J.L.: The Theory of Vibrational Spectroscopy and its Application to Polymeric Materials, New York: Wiley 1982
57. Henniker, J.C.: Infrared Spectroscopy of Industrial Polymers, London: Academic Press 1967
58. Neto, N., DiLauro, C.: Eur. Pol. J. *3*, 645 (1967)
59. Petcavich, R.J., Coleman, M.M.: J. Pol. Sci. Pol. Phys. ed. *18*, 2097 (1980)
60. Wu, M.S., Painter, P.C., Coleman, M.M.: J. Pol. Sci. Pol. Phys. ed. *18*, 95 (1980)
61. Simak, P.: J. Macromol. Sci. Chem. A *17*, 923 (1982)
62. Blais, P., Day, M., Wiles, P.M.: J. Appl. Pol. Sci. *17*, 1895 (1973)
63. Kato, K.: J. Appl. Pol. Sci. *19*, 951 (1975)
64. Sage, D., Berticat, P., Vallet, G.: Ang. Makrom. Chem. *54*, 151 (1976)
65. Holland-Moritz, K., Holland-Moritz, I., van Werden, K.: Coll. Pol. Sci. *259*, 156 (1981)
66. Holland-Moritz, K., Siesler, H.W.: Pol. Bull. *4*, 165 (1981)
67. Varnell, D.F., Runt, J.P., Coleman, M.M.: Macromolecules *14*, 1350 (1981)
68. Varnell, D.F., Coleman, M.M.: Polymer *22*, 1324 (1981)
69. Coleman, M.M., Sivy, G.T.: Carbon *19*, 123 (1981)

Part D. Spin-Resonance Spectroscopy

8 Principles of Spin-Resonance Spectroscopy

8.1 Introduction

The group of spectroscopic methods that are based on the spin-resonance effect has in common with those discussed in the previous chapters that it is due to true resonance absorption or emission of electromagnetic radiation, albeit under the additional action of external (laboratory) or internal (molecular) magnetic fields. Continuing the non-magnetic absorption spectroscopy beyond the fIR into the microwave region would lead to the relaxation spectra of polymers (the rotations typical of this spectral regions are irrelevant to macromolecules) which are due to slow movements of the polymer chains or parts of it. These non-resonance spectra, however, have been excluded from the scope of this book (see Part A).

The energy levels between which spin resonance absorption and emission takes place are in general created by external magnetic fields. Exceptions are the zero-field transitions of triplet states in ESR and the nuclear quadrupole magnetic resonance in NMR spectroscopy.

In Spin-Resonance Spectroscopy the interaction responsible for absorption of the low energy (microwave and radio frequency) is a magnetic one, whereas the interactions considered in the preceding chapters were nearly exclusively electric interactions. The field of Spin-Resonance Spectroscopy of polymers will be conveniently classified into Electron Spin-Resonance (ESR, Chap. 9) and Nuclear Magnetic Resonance (NMR, Chap. 10). Before discussing these methods in further detail, a few questions concerning the common basic principle of the two will be treated in this introductory chapter:

- What is an electron or a nuclear spin?
- How does resonance absorption come about in systems containing spin?
- Which polymers can be investigated by Spin-Resonance Spectroscopy?

8.2 The Spin of Elementary Particles

Spin is a movement which in the macroscopic world is known as rotation about an axis of a body (e.g. of a top). This movement is characterised by an angular momentum (8.1).

$$\text{Angular momentum} = \text{Moment of inertia } (\Theta) \times \text{Angular velocity } (\omega). \qquad (8.1)$$

The dimension of angular momentum is identical with that of the Planck constant, h (energy $\times$ time), the SI unit being $Js \equiv kg\, m^2\, s^{-1}$.

The elementary particles forming the atoms (e^-, p^+, n) are known to show an angular momentum of $\dfrac{1}{2} \cdot \dfrac{h}{2\pi} \equiv \hbar/2$. This elementary angular momentum, which in nuclei and atoms may add vectorially to give $0, 1/2, 1, 3/2 \ldots X \hbar$ is briefly called spin and characterised by the numbers $0, 1/2, 1, 3/2 \ldots X$ omitting $\hbar$. The angular momentum is the physical reality which has been postulated from quantum mechanics (Pauli-Dirac[1]) and experimentally verified. A spin-like movement of the particles corresponding to this angular momentum can be imagined, taking into account the tentative nature of all macroscopic pictures of atomic events.

The electrons in atoms and molecules may in addition to the above spin angular momentum show an orbital angular momentum which combines with the former to yield a total angular momentum. Ordinary ("closed shell") molecules mostly have a total electron spin of zero (singlets). Radicals have a total electron spin of $1/2$ corresponding to one unpaired electron (doublets), triplet states with two unpaired electrons have an electron spin of 1 (see Sect. 4.5.1).

In elementary particles, the angular momentum is furthermore connected with a magnetic (dipole) momentum, the vector $\boldsymbol{\mu}$. This can easily be understood in the case of charged particles, but much less so in the case of the neutron.

The magnetic moment of the free electron amounts to

$$\mu = 9.285 \times 10^{-24}\ JT^{-1}\ (\equiv A\, m^2) \tag{8.2}$$

and numerically nearly equals Bohr's magneton μ_B which is defined according to (8.3) by universal constants only:

$$\mu_B = \frac{eh}{4\pi m_e} = 9.274 \times 10^{-24}\ JT^{-1}. \tag{8.3}$$

Defining an analogous nuclear magneton μ_N [using the mass of the proton, m_P, instead of m_e in Eq. (8.3)] we see that his quantity has to be smaller by the factor $m_P/m_e = 1836.15$ than the corresponding electronic (Bohr's) moment.

$$\mu_N = \frac{eh}{4\pi m_p} = 5.0508 \times 10^{-27}\ JT^{-1}. \tag{8.4}$$

The experimental magnetic moment of the proton amounts to $2.79\ \mu_N$.

Summing up we note that the mechanic angular momentum – the spin – is connected with a magnetic moment. This magnetic moment is smaller roughly by a factor of 1,000 for the proton compared to the electron.

8.3 Resonance Absorption

In order to observe resonance absorption (or emission) of electromagnetic radiation, a magnetic field which acts on the elementary magnetic moments of the electrons or nuclei is necessary [3, 8–11]. The force of this interaction is proportional to the magnetic

Tabelle 8.1. Resonance Frequencies and g_N Factors of Several Nuclei, Relevant to Polymer NMR

Nucleus	Natural abundance percent	v_0 at 2.114 T MHz	g_N
^{1}H	99.98	90.00	5.5854
^{2}H(D)	1.6×10^{-2}	13.82	0.8574
^{13}C	1.1	22.63	1.4044
^{14}N	99.64	6.50	0.4036
^{19}F	100	84.67	5.2546
^{33}S	0.74	6.90	0.4284
^{31}P	100	36.43	2.2610

flux density **B** of the field [1] and to the magnetic moment of the particle. The resonance condition in its general form is given in Eq. (8.5).

$$hv = g\mu B \tag{8.5}$$

g = proportionality factor ($g_e = 2.0$ in case of free electrons, $g_N = 0.1$ to 6 for many nuclei, see Table 8.1); $\mu = \mu_B$ for electrons, μ_N for nuclei; $B = |\mathbf{B}|$, magnetic flux density.

Figure 8.1 shows that part of the electron-magnetic spectrum which is used in Spin-Resonance (ESR and NMR) Spectroscopy.

According to this scheme, ESR spectroscopy is performed in the microwave region, around 10 GHz, whereas NMR requires short wavelength radiowaves in the 1 to 10-m band. The magnetic strength required is of the order of 1 Tesla (10^4 Gauß). Usually, the sample is placed into a strong, homogeneous magnetic field and exposed to microwaves of RF radiation. In order to obtain a spectrum, the frequency could in principle be scanned under fixed B [Eq. (8.5)] over the absorption region. This procedure would correspond to the use of monochromators in optical spectroscopy. Experimentally, however, it is easier to scan the magnetic flux density B, while keeping the frequency constant. Furthermore, in ESR spectroscopy, the first derivative of the absorption dA/dB is recorded as a function of B rather than the absorption A itself.

The first ESR spectra have been obtained by Savoisky in 1945[4]. The NMR technique was introduced in the same year by Purcell and Bloch[5] and has been applied to the study of polymers soon after its discovery[6].

8.4 Spin-Resonance in Polymers

Most polymers are diamagnetic, thus containing only spin-compensated electrons. Such polymers can only be investigated by ESR if excited to paramagnetic triplet states (see Sect. 4.5.1), or if unpaired electrons are introduced such as in
- polymers containing "broken bonds", e.g. by mechanical stress[12] or UV radiation
- polymers produced as polyradicals[7]

---

1 The magnetic flux density is proportional to the magnetic field strengh **H** according to $B = \mu_0 \mathbf{H}$[2] where $\mu_0 = 4\pi \ 10^{-7}$ Vs/Am is the permeability of vacuum (SI units)

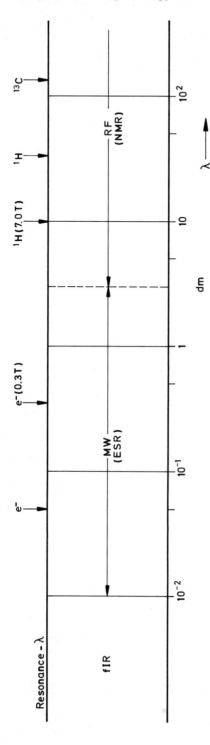

Fig. 8.1. Wavelength Region of Spin-Resonance Spectroscopy. The resonance wavelengths are given for $B = 2.114\ T$ unless indicated otherwise

– polymers caused to react with "spin labels"
– polymers containing transition metal ions, e.g. biopolymers.

NMR can be applied to several nuclei usually present in polymers (^{1}H, ^{13}C), although in many cases g_N is small, resulting in low intrinsic sensitivity and the natural abundance of the active nucleus may be small (see Table 8.1). In these cases, special techniques (pulse-FT NMR) have to be applied. Only ^{19}F equals ^{1}H with regard to g_N and natural abundance of the magnetically active isotope. The increase in sensitivity at high g_N is due to the proportionally higher energy difference [see Eq. (8.5)] between ground and excited states at fixed B and temperature, thus yielding a favourable Boltzmann distribution between ground and excited states. The natural abundance does, of course, influence the concentration of the particles available for detection at a given concentration of the polymer to be studied. In contrast to the more specific ESR technique, NMR spectroscopy can be applied nearly universally in polymer research. In this regard it may be compared with mIR absorption spectroscopy.

References

1. Schpolski, E.W.: Atomphysik, I and II. VEB Berlin (1973/1974)
2. Pohl, R.W.: Elektrizitätslehre, 20. Auflage. Berlin, Heidelberg, New York: Springer 1967
3. Ingram, D.J.E.: Spectroscopy at Radio and Microwave Frequencies, 2nd Ed. London: Butterworth 1967
4. Zavoisky, E.: J. Phys. USSR 9, 211 (1945); see also: B. Rånby in: ESR Applications to Polymer Research, Nobel Symposium 22. Stockholm: Almquist & Wiksell 1973
5. Bovey, F.A.: High Resolution NMR of Macromolecules. New York: Academic Press 1972
6. Slichter, W.P.: Fortschr. Hochpolymer Forsch. 1, 35 (1958)
7. Braun, D.: J. Pol. Sci. Part C 24, 7 (1968)
8. Mc Lauchlan, K.A.: Magnetic Resonance. Oxford: University Press 1975
9. Coogan, C.K., Ham, N.S., Stuart, S.N., Pilbrow, J.R., Wilson, G.: Magnetic Resonance. New York: Plenum 1970
10. Erbeia, A.: Resonance Magnétique. Paris: Masson 1969
11. Poole, C.P., Farah, H.A., Jr.: The Theory of Magnetic Resonance. New York: Wiley-Interscience 1972
12. Kausch, H.H.: Polymer Fracture. Berlin, Heidelberg, New York: Springer 1978

9 Electron-Spin-Resonance (ESR) Spectroscopy of Polymers

9.1 General Characteristics of ESR Spectra

The basis of ESR spectroscopy is absorption of electromagnetic radiation by unpaired electrons in a magnetic field [1, 5–9]. Unpaired electrons are characterised by an angular momentum, the "spin", which is connected with a magnetic moment (Chap. 8). This magnetic moment is influenced by external magnetic fields in such a way that two orientations of different energy are formed [Eq. (9.1) and Fig. 9.1].

Without external field, all unpaired electrons are degenerate, i.e. they have equal energies. It is therefore permissible to say that the external magnetic field removes this degeneracy. Since the degenerate energy levels of a single unpaired electron split into two, the corresponding state is called a doublet in analogy to optical spectroscopy of

D. Spin-Resonance Spectroscopy

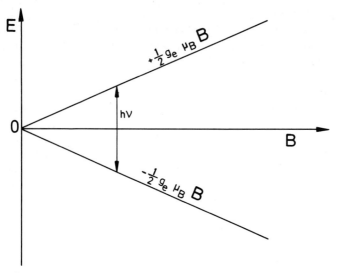

Fig. 9.1. Schematic presentation of the splitting of energy levels in a magnetic field (doublet). B: magnetic flux density; see Eq. (9.1)

atoms and small molecules in the gas phase where similar splittings are observed (coupling of spin transitions with electronic transitions).

Let us consider a polymeric radical, e.g. a broken C—C chain:

$$\text{wwwC} \overset{\displaystyle {}^1\text{H}}{\underset{\displaystyle {}^1\text{H}}{\overset{|}{\underset{|}{-}}}} {}^{12}\text{C}\bullet$$

In a first approximation, we expect a resonance absorption close to that observed for free electrons; since the magnetic interaction of un unpaired electron with its surroundings is in general weak:

$$h\nu = g_e \mu_B B \tag{9.1}$$

$g_e = 2.0023$ (free electron "g-factor"), $\mu_B =$ Bohr's magneton (see Chap. 8).

In contrast to most other spectroscopic techniques the resonance frequency of ESR depends not only on molecular properties, but primarily on the external magnetic field, as measured by the magnetic flux density B [Eq. (9.1)]. For experimental reasons, in ESR spectroscopy two microwave frequencies, called X band and Q band (Table 9.1), are preferentially used (as indicated by the term "band", X and Q indicate a range of frequencies available).

Since $\Delta E = h\nu$ is very small ($1.6 \times 10^{-3} kT$ at room temperature in the case of X-band), the upper and the lower energy levels are nearly identically populated so that "upward" and "downward" transitions are almost equally likely to occur if microwaves interact with the sample, which is located between the poles of a strong magnet of variable field strength. The net absorption observed is due to the slightly superior population of the lower "ground state" [see Eq. (9.5)]. During the experiment, ν is kept con-

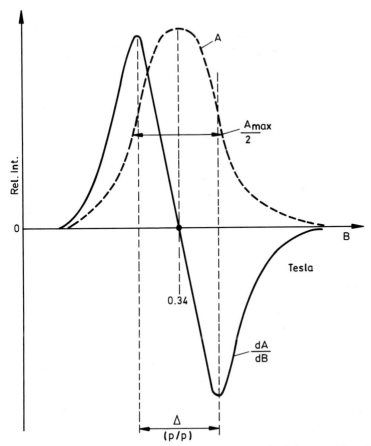

Fig. 9.2. Schematic ESR absorption spectrum of an unpaired electron without hyperfine splitting; absorption (A) in relative units and first derivative dA/dB. The position of maximum absorption is given for X-band

stant. Scanning B through the resonance region according to (9.1), we expect a roughly Gaußian absorption surve, $A(B)$, as shown in Fig. 9.2.

Again for experimental reasons, the 1st derivative of the absorption curve, $dA/dB(B)$ is recorded instead of $A(B)$. A low frequency alternating field is superposed on the – slowly scanned – external magnetic field and the resonance absorption thus obtained is amplified and recorded to yield directly the derivative absorption dA/dB. As a measure of line-width, the easily derived Δ peak-to-peak width (see Fig. 9.1) is used instead of Δ at $A_{max}/2$ ("half-width") which is somewhat larger. The minimum line-width is given by the uncertainty principle (9.2).

$$\Delta_{min} \times \Delta t \approx \hbar/2, \tag{9.2}$$

$$\Delta_{min}(\text{energy}) = h\Delta v = g_e \mu_B \Delta B. \tag{9.3}$$

If Δt is extremely small, i.e. if the relaxation is very fast, Δ may become large relative to ordinary ESR band widths and the ESR absorption may actually vanish.

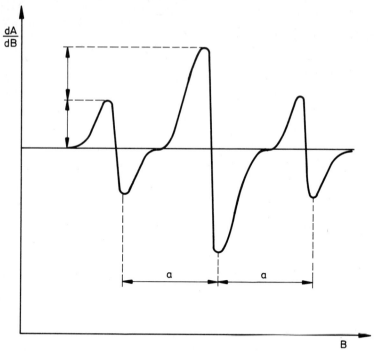

Fig. 9.3. ESR spectrum of an unpaired electron coupled with two equivalent protons; a = coupling constant

According to Eqs. (9.2) and (9.3), a natural line-width of

$$\Delta B = 1 \text{ G} (10^{-4} \text{ T})$$

corresponds to a relaxation time of about 30 ns. Compared to other spectroscopic methods, we observe in ESR (and even more so in NMR) extremely narrow lines and as a consequence the natural line-width may influence the observed line shape in many cases. The relaxation processes deactivating the energetically higher ("excited") state created by absorption of a microwave photon are

Spin-lattice relaxation: $\Delta t = T_1$

and

Spin-spin relaxation: $\Delta t = T_2$.

The relaxations are relatively slow processes in C-radicals and similar unpaired electrons in organic molecules, e.g. in polymer radicals and thus in general do not strongly increase Δ in ESR spectra.

In transition metal ions the relaxation may be very fast due to spin-orbit coupling, so that ESR may often be observed only at very low temperatures. This point is crucial when recording ESR spectra of some biopolymers [2].

A typical ESR spectrum is shown in Fig. 9.3. This spectrum could be due to the above $\sim\sim$CH$_2 \cdot$ radical, taking now into account the hyperfine coupling between the

116

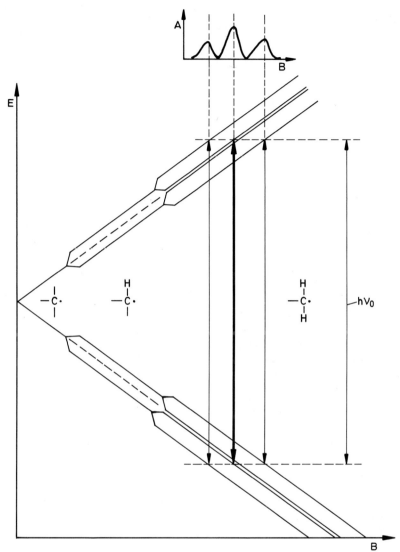

Fig. 9.4. Pictorial explanation of hyperfine splitting (after Rånby and Rabek [1]; $h\nu_0$ indicates the microwave energy used

unpaired electron and the two equivalent protons (the actual recording of this particular spectrum depends, of course, on the strict exclusion of oxygen). Figure 9.3 shows three peaks, the central peak being twice as strong as the outer peaks, which have the same distance (a) from the central peak. The true absorption curve A (B) (Fig. 9.4) would give the same information. The reason for this splitting is the "isotropic" (i.e. independent of the direction of the magnetic field) interaction of the magnetic moments of the protons of the radical end group.

Owing to this coupling, the electron and nuclear spins orient one other inducing small energy differences and thus slightly different resonance frequencies. In a pictorial

117

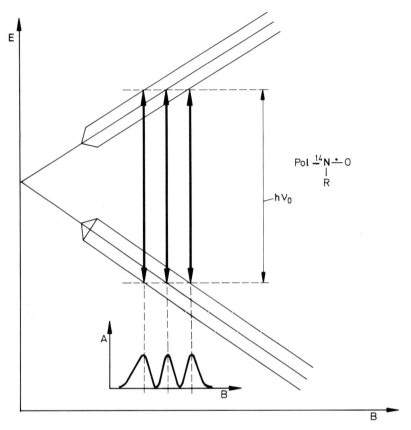

Fig. 9.5. Hyperfine splitting between an unpaired electron and a nucleus with spin 1 (e.g. a spin label of the type indicated). As in Fig. 9.4, splitting is strongly exaggerated for the sake of clarity [1]

way, this splitting, first by one proton, then by a second one, is shown in Fig. 9.4. Using a fixed microwave frequency v_0, the experimental absorption spectrum can easily be deduced from this diagram (insert in Fig. 9.4). Interaction of the unpaired electron with only *one* proton would result in a doublet of equally intense ESR absorption lines (inner part of Fig. 9.4). The (isotropic) coupling constant a depends on the unpaired electron density at the protons or, generally, at the interacting paramagnetic nuclei.

In organic molecules, ^{14}N is a further important nucleus [1], because of its high natural abundance and its nuclear spin 1 corresponding to two parallel elementary spins, ("nuclear triplet state"). In polymer spectroscopy, ^{14}N splitting is observed, e.g. in spin labels. As can be seen in Fig. 9.5, splitting leads to three equally intense hyperfine structure lines.

The simultaneous interaction of an unpaired electron with more than one or two paramagnetic nuclei leads to intricate spectra with many lines except for highly symmetrical molecules. The coupling constants (a) can be deduced from measured peak distances and thus a picture of the electron density distribution within the molecule or "chromophore" – e.g. a side group of an aromatic polymer can be obtained. In this case, the spectral resolution has to be high, a requirement which is achieved only with difficulty in polymers.

The density of the unpaired electron (ϱ) at the interacting nucleus or at a neighbouring nucleus (e.g. H splitting in aromatic radical-ions) is porportional to the experimental hyperfine splitting constant, a, Eq. (9.4).

$$a_H = Q_H \varrho .$$ (9.4)

$Q_H = 22.5$ G in $C_6H_6^{\cdot -}$; similar values have been found for other aromatic radical ions [3]; in this case is the density of the unpaired electron obtained, e.g. from HMO calculation at the C-atom to which the interacting H-atom is connected.

In polymers, radical anions (e.g. of P1VN) tend to degrade so that ESR spectra have to be interpreted with care [4].

As indicated above, line broadening, especially of polymers, is a major problem in ESR spectroscopy. This may be due to
– extremely fast relaxation (T_1 and/or T_2 very short)
– saturation of signals due to high microwave power
– anisotropic hyperfine interaction or other anisotropic line broadening effects.

Fast relaxation leads to homogeneous line broadening, i.e. the line is broadened as a whole. From this point of view, slow relaxation seems to be desirable, although this leads to another broadening mechanism – signal saturation – which is due to equal distribution of lower (N_0) and higher (N_s) energy radicals. Without perturbation by a microwave field the N_s/N_0 ratio is determined by the Boltzmann distribution (9.5),

$$\frac{N_s}{N_0} = e^{-\frac{\Delta E}{kT}} = e^{-\frac{g_e \mu_B B}{kT}} .$$ (9.5)

For X band frequencies and $T = 295$ K, we obtain $N_s/N_0 = 0.985$.

Since resonance interaction with the microwave field leads to absorption and induced emission of photons with equal probability, net absorption is not observed if $N_s/N_0 = 1$. The intensity of the ESR signal increases with microwave power; in slowly relaxing systems, however, this effect cannot be fully used to increase the sensitivity because higher power levels saturate ($N_s/N_0 \approx 1$) the ESR signal.

For polymers, as for all amorphous solids and highly viscous solutions, anisotropic or "classical" interaction of the unpaired electron with surrounding paramagnetic nuclei or other unpaired electrons is an important line-broadening mechanism. This broadening is due to the small additional magnetic fields provided by the paramagnetic particles (nuclei or unpaired electrons) which increase or decrease the external magnetic field by small amounts and thus are shifting the resonance flux density B_0 over a certain region, the width of which determines the anisotropic broadening (Fig. 9.6).

Whereas in fluid solutions the magnetic dipolar interactions are averaged to zero, in amorphous and polycrystalline solids and in viscous solutions, each unpaired electron "feels" a somewhat different magnetic field. The broadening due to this effect is furthermore inhomogeneous, since the signal observed is composed of many different lines, each corresponding to a different resonance frequency of B_0. This can be shown directly by partial saturation of one of the lines, which causes a dip in the absorption signal ("hole burning"). Rigid polymer molecules can show anisotropic interaction

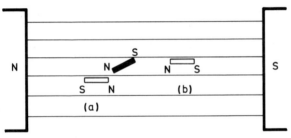

Fig. 9.6. Visualisation of anisotropic line broadening of an ESR active spin, depicted as small magnet (black bar) by other nearby elementary magnets (white bars); (*a*) increase of external field; (*b*) decrease of external field

even in fluid solution unless the frequency of molecular movements > hyperfine splitting (a splitting of 10 G corresponds to 28 MHz).

A last type of band broadening is due to the small, but measurable anisotropy of g_e [Eq. (9.1)]. The average φ_e, as measured in liquid solutions is mostly found near the value of the free electron (2.0023). Small deviations from this value can be used analytically in order to identify radicals. In single crystals of known orientation relative to the magnetic field, the anisotropy of g_e can be measured; in amorphous and polycrystalline media where the individual spins do not have a common orientation, this anisotropy causes line broadening and distortions.

9.2 Experimental

An ESR spectrometer consists of the following essential parts [1, 2, 5−9], see Fig. 9.7. The radiation source is a klystron in which a reflected electron beam is modulated by means of electric fields thereby producing monochromatic microwaves of a relatively small power of about 1 to 100 mW. The frequency chosen can be precisely kept constant by modifying the tension of the reflector (AFC). The modulation range of a klystron is about $v_0 \pm 25$ MHz at $v_0 \approx 10$ GHz (X band). A monochromator is not necessary owing to the purity of the frequency generated.

The microwaves are conducted in a suitably formed metallic hollow waveguide to the cacity to which they are coupled by means of a hole (iris). The cavity is designed in the form of a resonator, in which standing microwaves are formed by reflection and interference (Fig. 9.8). The sample has to be introduced into the cavity in the region of the highest magnetic field strength (interaction with the magnetic moments) which corresponds to the minimum electric field strength so that dielectric losses in the sample are minimised. This is especially important when water is used as solvent; in that case, flat absorption cells, parallel to the xy plane (Fig. 9.8) have to be used. The resonator has to be tuned exactly to the frequency used. The condition of interference requires the dimensions of the cavity to be of the order of the wavelength (Table 9.1).

Part of the difficulties in working with Q-band frequencies is due to the smaller sample – and cavity dimensions required compared to those for X-band frequencies. The magnet has to produce a maximum magnetic flux density of the order of $B_{max} = 1$ Tesla and has to be variable over the whole region from $B = 0$ up to B. The field has to

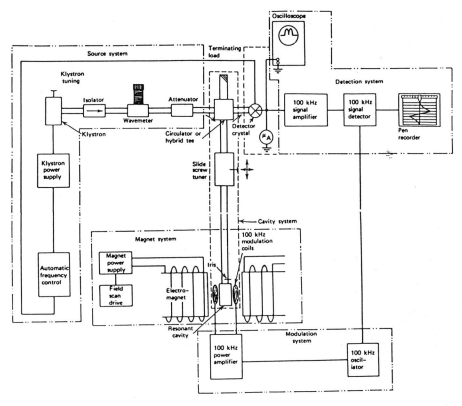

Fig. 9.7. Block diagram of an X-band ESR spectrometer equipped with 100 kHz phase – sensitive detection (after [2])

be homogeneous and constant within ± 1 ppm. Most modern ESR spectrometers use 100 kHz modulation of the magnetic field produced by small, additional coils located at the cavity walls. Thus, periodical changes in ESR absorption (A) are created, which are used for the phase-sensitive amplification of the signal (Fig. 9.9).

Recording very narrow peaks, it has to be taken into account that

$$v \text{ (modulation)} < \Delta \qquad\qquad (9.6)$$

is required; in X band, 100 kHz corresponds to about $3\,\mu T$ (30-mG); polymer peaks, however, are mostly much broader.

The detector (Si crystal), which is located outside the cavity in a branch of the waveguide, is sensitive to the weakening of the microwave field by the sample. The 100 kHz component is amplified selectively and either recorded directly or stored to improve the S/N ratio using signal averaging. Strong signals can immediately be observed by an oscilloscope.

Under favourable conditions meaning small line-width $\Delta = 1$ G and insignificant saturation tendency so that high MW power can be used, the sensitivity is reported to

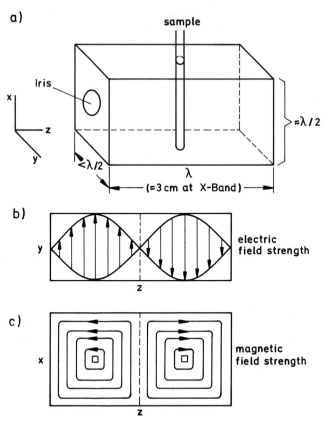

Fig. 9.8 a–c. Scheme of a TE_{102}-resonator cavity **(a)**, electric **(b)**, and magnetic **(c)** field distribution showing maximum magnetic field strength at the sample position

Table 9.1

X-Band	Q-Band
$v\ =9.5\ \text{GHz}$	$v\ =35\ \text{GHz}$
$\lambda\ =3.16\ \text{cm}$	$\lambda\ =\ 8.57\ \text{mm}$
$B_X=0.339\ \text{T}$	$B_Q=\ 1.25\ \text{T}$

be near 10^{10} "spins" [1] [28)] per sample.

10^{10} "spins" correspond roughly to

10^{-2} pmol (absolute) or 10^{-10} mol l^{-1}.

The concentration can be determined, albeit not very exactly, by comparison with standard samples. In modern ESR spectrometers, on line-computers are used for data treatment and simulation of spectra [28)].

---

1 In laboratory jargon, "spin" is used synonymously for "unpaired electron"

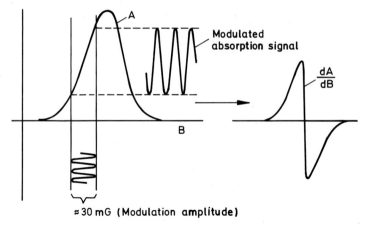

Fig. 9.9. Modulation of ESR absorption. The intensity of the modulated signal depends on dA/dB

9.3 Survey of Polymer-Specific Applications of ESR Spectroscopy

According to Rånby [28], the first applications of ESR spectroscopy to polymers have been reported by Bresler [29], Bamford [30], and Ingram [31] in the years 1958 to 1960. Since that time, a large number of papers have been appearing.

The most comprehensive monography on this topic is the book by Rånby and Rabek [1]. Special fields have been treated, e.g. by Fischer [10], Kausch [11], Tormala and Lindberg [12] and many others. Biopolymers have been treated by Keighley [13].

According to these sources, the main fields of polymer-related applications for ESR are:

– Radicalic polymerisation
– Mechanical degradation and fracture of polymers
– Radiation- and light-induced radical formation
– Spin label and spin probe techniques
– "True" electron-spin polymers
– Conductive polymers
– Triplet states and biradicals.

In the following sections, a few examples will be given for the most important applications. Since ordinary polymers are mostly diamagnetic, ESR is used for studying growing polymer chains, degradation processes and optically excited or chemically modified polymers.

9.4 Polymerisation Studies by Means of ESR

For a large number of polymerisation processes the following mechanism had been postulated, long before ESR was used for the direct observation of radicalic intermedi-

123

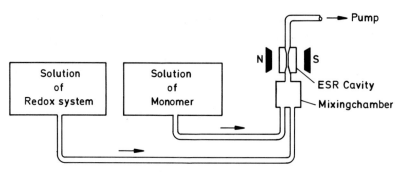

Fig. 9.10. Scheme of a flow system for ESR studies of radicalic polymerisation

ates:

Initiator $\xrightarrow[h\nu]{\Delta \text{ or}}$ Starting radical

$+$ Monomer $\longrightarrow$ Monomeric radical $(M\cdot)$

$+$ Monomer $\longrightarrow$ Dimeric radical $\quad (D\cdot)$

$+$ Monomer $\longrightarrow$ Polyradical $\quad\quad (P\cdot)$

Termination (recombination, chain transfer, etc.)

$\longrightarrow$ Saturated macromolecule
(polymer or oligomer).

The experimental problem involved in the direct detection of the intermediate radicals is their high reactivity and, hence, low stationary concentration. Figure 9.10 shows a flow system which can be used to generate a high concentration of radicals in the cavity of the ESR spectrometer, using Fenton's reagent (9.7) or the corresponding Ti^{3+} system (9.8) as initiator (OH as starting radical if H_2O is used as solvent) [10, 14)

$$Fe^{2+} + H_2O_2 \rightarrow Fe^{3+} + OH^- + OH, \tag{9.7}$$

$$Ti^{3+} + H_2O_2 \rightarrow Ti^{4+} + OH^- + OH. \tag{9.8}$$

According to the general scheme, the reaction proceeds as follows (9.9):

$$
\bullet OH + CH_2{=}\underset{\underset{R_2}{|}}{\overset{\overset{R_1}{|}}{C}} \rightarrow HO{-}CH_2{-}\underset{\underset{R_2}{|}}{\overset{\overset{R_1}{|}}{C}}{\bullet} \rightarrow (D^\bullet) \rightarrow
$$

Starting radical $\quad\quad$ Monomer $\quad$ Monomeric radical $(M\cdot)$

$$
HO{-}CH_2{-}\underset{\underset{R_2}{|}}{\overset{\overset{R_1}{|}}{C}}\left[{-}CH_2{-}\underset{\underset{R_2}{|}}{\overset{\overset{R_1}{|}}{C}}{-}\right]CH_2{-}\underset{\underset{R_2}{|}}{\overset{\overset{R_1}{|}}{C}}{\bullet} \tag{9.9}
$$

Poly$-$radical $(P^\bullet)$

124

Table 9.2. Hyperfine splitting parameters for polymerizing acrylic acid

Radical		$a(H_\alpha)$ G	$a(H_\beta)$ G
$M^\bullet$	$HO-\overset{\beta}{C}H_2-\overset{\alpha}{\underset{\underset{COOH}{\mid}}{\overset{\overset{H}{\mid}}{C}}}{}^\bullet$	20.23	22.81
$D^\bullet$ ($\approx P^\bullet$)	$HO-CH_2-\overset{\overset{H}{\mid}}{\underset{\underset{COOH}{\mid}}{C}}-\overset{\beta}{C}H_2-\overset{\alpha}{\underset{\underset{COOH}{\mid}}{\overset{\overset{H}{\mid}}{C}}}{}^\bullet$	22.62	21.34

Other starting radicals are $^\bullet CH_2OH$ (using methanol as solvent), CH_3 (tert. butylhydroperoxide instead of H_2O_2) or $^\bullet NH_2$ (from NH_2OH).

Direct evidence of the radicalic mechanism of polymerisation is obtained by detecting radicalic intermediates, which is possible by ESR. The intermediates are identified using the hyperfine parameters, or differences in g_e values of $M^\bullet$, $D^\bullet$ and $P^\bullet$. The radicals expected according (9.9) can indeed be identified by ESR. Using vinyl monomers ($R_1 = H$), it was shown that the growing end of ($P^\bullet$) is the "head" [as shown in (9.9)] rather than the "tail". This behaviour had earlier been postulated on the basis of indirect evidence.

The radicals ($P^\bullet$) $\approx$ ($D^\bullet$) and ($M^\bullet$) are chemically very similar so that high resolution is needed in order to resolve the spectra as shown in Table 9.2 for acrylic acid. The similarity of the hyperfine coupling constants a ($H\alpha$) and ($H\beta$) indicates a similar spin density at the two proton positions. If NH_2 was the starting radical, even a weak ^{14}N triplet can be observed in $M^\bullet$ ($a(N) = 3.4$ G), but not in $D^\bullet$ and $P^\bullet$, thus labelling the monomer radical.

Kinetic experiments can be performed using different concentrations of [M] and following up the radicals quantitatively in order to elucidate the elementary steps of polymerisation.

Other applications of ESR to polymerisation processes include:

− Copolymerisation:
  the ESR spectrum shows which monomer is added preferentially to the growing chain end, e.g. in the copolymerisation of crotonic compounds with acrylonitrile [32].
− Anionic polymerisation:
  the growing end of a "living polymer" in a purely anionic polymerisation is diamagnetic, e.g.

$$\left[\sim CH_2-\overset{\overset{R_1}{\mid}}{\underset{\underset{R_2}{\mid}}{C}}{}^\ominus \right] Me^+$$

and thus not detected by ESR; precursors and intermediates of the actual initiator, such as naphthalene$^{\cdot-}$ can easily be studied by ESR.

– Cationic polymerisation:
for this type of polymerisation, if pure, the same is true as for the anionic polymerisation. Again, precursors can be identified, e.g. CT complexes forming the initiator by light-induced electron transfer.

– Solid-state polymerisation:
in favourable cases, allowing a fixed orientation of the growing radical with regard to the external magnetic field, this type of polymerisation may be carried out in single crystals. Often, however, the crystal lattice will be destroyed during polymerisation, introducing incertainties in the interpretation of spectra. The initiation by means of ionising radiation may furthermore create radicals not involved in starting and propagating the chain growth! See also Section 9.7.

– Radiation – induced ionic polymerisation [15]:
in the solid state or in solid matrix. The intermediates (solvent$^{\cdot+}$, $M^{\cdot-}$ and/or $M^{\cdot+}$) can be detected using ESR. The primary products in rigid solution are the radical cation of the solvent molecules and free electrons.

9.5 Mechanically Formed Polymer Radicals

Methods for producing "mechanoradicals" include [11]:

a) Milling, cracking, sawing to give powders of large surface area and high radical concentration.

b) Stretching of fibres in the cavity; this method produces fewer radicals than (a) but oriented ones, and their formation can be recorded during stress.

The ESR spectra observed originate from three types of radicals:

(1) Primary radicals – "broken chains" – such as $\sim\sim CH_2$; these radicals are extremely reactive and are converted into

(2) Secondary radicals. In polyamides (PA), the secondary radical which is relatively stable in the absence of O_2 is invariably $^{\cdot}CH$ next to the amide nitrogen (9.10).

$$\sim\sim CH_2 \cdot \quad
\begin{array}{c} \zeta \\ CH_2 \\ | \\ H-C-H \\ | \\ NH \\ | \\ C=O \\ | \\ PA \end{array}
\longrightarrow
\begin{array}{c} \zeta \\ CH_2 \\ | \\ \bullet C-H \\ | \\ NH \\ | \\ C=O \\ | \\ PA \end{array}
\qquad (9.10)$$

Primary radical   Secondary radical (I)

This radical (I) is formed by hydrogen transfer. The localisation of the unpaired electron in the C(H)—N results in the absence of accidental degeneracies in 3 (N) × 2 (H) ESR absorption lines (Fig. 9.11).

(3) Follow-up radicals which are due to chemical reactions of the primary or secondary radicals with impurities. The most important radical of this kind – under practical

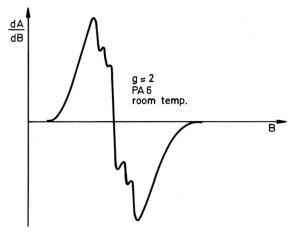

Fig. 9.11. ESR spectrum of the metastable secondary radical of PA 6, see Eq. (9.10) at room temperature; $g_e \approx 2$

conditions of incomplete exclusion of oxygen – is the peroxy radical:

$$H-\overset{\wr}{\underset{\wr}{C}}\bullet \; + \; O_2 \;\longrightarrow\; H-\overset{\wr}{\underset{\wr}{C}}-O-O\bullet$$

The unpaired electron in the peroxy radical does not have an appreciable density at any paramagnetic nucleus and therefore forms a single ESR line which may be distorted by an anisotropy of the g_e-factor. During fracture under "ordinary" environmental conditions (e.g. $[O_2] \approx 10^{-3}$ mol l^{-1}, room temperature), the three types of radicals appear at the same time. In ESR experiments aimed at elucidating the mechanism of fracture and radical formation, an attempt is made to separate the stages, measuring gradually, first at low temperature (77 K), warming up, adding oxygen etc. The spectra are poorly resolved in most cases so that additional information (irradiation experiments, etc.) are often needed in order to definitely assign the splittings observed.

In PA-6[11], the primary radicals at 77 K are distributed as follows (9.11):

$$\begin{aligned} &50\% \quad \sim CH_2-CH_2^{\bullet} \quad (A) \\ &25\% \quad \sim\overset{\textstyle |}{\underset{\textstyle O}{C}}-CH_2^{\bullet} \quad (B) \\ &25\% \quad \sim\overset{\textstyle |}{\underset{\textstyle H}{N}}-CH_2^{\bullet} \quad (C). \end{aligned} \tag{9.11}$$

This distribution shows clearly that primary breaking occurs with equal probability at the α-β-bond at both sides of the amide groups.

$$\sim\!\!\sim\!\!N-\underset{\overset{|}{H}}{\overset{|}{\underset{}{C}}}-CH_2\!\!\mathrel{\underset{\alpha}{\text{---}}}\!\!CH_2-CH_2-CH_2\!\!\mathrel{\underset{}{\text{---}}}\!\!CH_2-N-C\!\!\sim\!\!\sim$$

PA–6

127

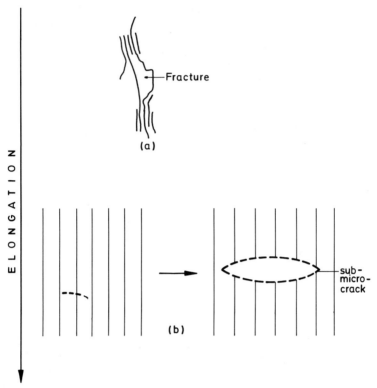

Fig. 9.12 a, b. Primary processes of fracture; Peterlin-Sohma **(a)** and Zhurkov-Kausch **(b)** models[11]

Warming up under strict exclusion of O_2 gives rise to the following secondary reactions [(9.12); I see Eq. (9.10)] without loss of total ESR-intensity,

$$120 \text{ K:} \quad A \rightarrow I$$
$$120\text{--}240 \text{ K:} \quad B + C \rightarrow I \tag{9.12}$$
$$> 240 \text{ K:} \quad I \text{ is the predominant radical.}$$

Stretching experiments using ESR during stress up to the beginning of fracture[1, 11, 16] are aimed at the pinpointing of the primary processes of fracture in an early stage where macroscopic damage canot yet be discerned. According to these studies fracture starts at single broken bonds and first leads to "submicro cracks"[16] which can also be identified using X-ray small angle scattering. For partly crystalline polymers the Peterlin-Sohma model is suitable which assumes strained macromolecules belonging to more than one microcrystal as the initial point of attack (Fig. 9.12a). The Zhurkov-Kausch model (Fig. 9.12 b), on the other hand, assumes the formation of more than one broken chain for each initially split macromolecule[11]. The primary splitting is followed by secondary reactions of the type Eq. (9.10) leading to saturated chain ends and macro radicals [e.g. I in (9.10)] which are more likely to break near the unpaired electron (weaker, electron deficient bonds) than elsewhere in the macromolecule.

128

It can be concluded from ESR studies of polymer fracture under elongation that most polymers do break radically and that the primary event is invariably a main chain splitting (rather than "stripping off" of side groups).

9.6 ESR of Radicals Formed by Radiation

In principle, any radiation whose energy per photon or particle is higher than the binding energy of the chemical bonds to be broken is capable of forming radicals (100 to 500 kJ mol^{-1} for most single bonds), provided that the radiation is absorbed by the polymer. In practice, γ, β and X-rays are used predominantly. For studies of the photolytic break-down and the photo-oxidation of polymers, nUV radiation is also used, ESR studies of polymer radicals formed by radiation are important with regard to the degradation of plastics and their stabilisation against radiation.

As discussed for the mechanoradicals, radiation – induced radicals are rapidly converted in the presence of oxygen to form peroxy radicals, hydroperoxides, ketones, etc. In the absence of oxygen, the primary reactions can be studied by ESR.

Polyolefines (RH) react with radiation predominantly according to Eq. (9.13)[1].

$$RH \xrightarrow{\gamma} RH^{\bullet+} + e^- \xrightarrow{\text{Recombination}} (RH)^* \longrightarrow R^\bullet + H^\bullet \quad (9.13)$$

Using n-alkane single crystal as oligometic models for PE it has been established by ESR that all C—H bonds are attacked with equal probability by the high energy photons; in a secondary reaction, the terminal —CH_2 groups are converted into —$\dot{C}H$-radicals. A radical which has been frequently observed in polyolefines is the metastable allyl radical.

$$\text{\small{www}—CH_2—CH = CH—\underset{\bullet}{\overset{H}{C}}—CH_2—\text{www}} \ .$$

The absorption of free electrons (e^-), to be expected according to Eq. (9.13), has indeed been identified as a sharp single peak at $g_e = 2.002$.

The literature on radiation-induced radicals in polymers is abundant [1, 18]. The case of PIB is chosen as an example to explain differences caused by different modes of radiation [1, 17].

The ESR spectrum of PIB which has been treated with ionising radiation at 77 K shows a doublet (two equally intense peaks, separated by $a_H = 20$ G) which is due to the radical (9.14).

$$\begin{array}{ccccc} & CH_3 & H & CH_3 & \\ & | & | & | & \\ \text{wwC}& & \underset{\bullet}{C} & & C\text{ww} \\ & | & & | & \\ & CH_3 & & CH_3 & \end{array} \quad (9.14)$$

PIB

UV irradiation at 77 K, on the other hand, gives first rise to a complicated ESR spectrum consisting of the superposed components A and B, as shown in Fig. 9.13. The component B, which is due to the $\dot{C}H_3$ radical vanishes after a few hours even at 77 K.

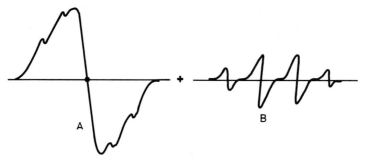

Fig. 9.13. ESR components of UV irradiated PIB, which, if superposed, forms the original spectrum at 77 K [1,17]

The broad band A is still a superposition of the signals originating from three radicals which are stable at 77 K,

(a) $-\overset{\bullet}{\underset{|}{C}}-$ (the counterpart of $^{\bullet}CH_3$)
  CH_3

(b) The radical (9.14)

(c) $\sim\!\!\sim\!\!CH_2-\overset{\overset{\displaystyle CH_3}{|}}{\underset{\underset{\displaystyle \overset{\bullet}{C}H_2}{|}}{C}}-CH_2\!\!\sim\!\!\sim$

Many plastics contain stabilisers against UV radiation (out-door use) and elevated temperature (during processing), one of the main stabilising mechanisms being the conversion of reactive radicals into more stable ones, e.g. hindered phenoxy radicals:

$\bullet O-\langle\!\!\!\!\!\!\!\bigcirc\!\!\!\!\!\!\!\rangle-R$

These reactions are conveniently studied using ESR spectroscopy [1].

9.7 Triplet States

Triplets are electronic states of two parallel spins ($\uparrow\uparrow$) which have been identified as electronically excited phosphorescent states in Sect. 4.5.1. Disregarding for the time being the magnetic interaction of the two spins, the energy of the triplet splits into three sublevels (hence the name "triplet") if a magnetic field is applied to the sample (Fig. 9.14). In this approximation, the triplet energy is given by Eq. (9.15) where the spin quantum number $m_s = +1, 0, -1$ (in contrast to $m_s = \pm 1/2$ for unpaired electrons, resulting in doublet states).

$$E_T = g_{el}\mu_B B m_s.\tag{9.15}$$

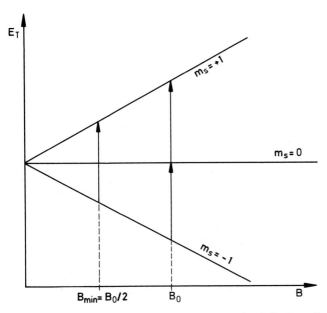

Fig. 9.14. Idealised (no zero field splitting) energy level diagram of a triplet system as a function of magnetic flux density B. The length of the arrows indicates the microwave photon energy which is kept constant in typical ESR experiments

Applying the selection rule $\Delta m_s = 1$, one ESR absorption is expected at B_0, roughly corresponding to the resonance absorption of free electron and a weak "half-field peak" which is due to the forbidden $\Delta m_s = 2$ transition (Fig. 9.14).

This oversimplified picture has to be modified for real systems taking into account the magnetic spin-spin interaction which causes zero field splittings (Fig. 9.15): the triplet states of molecules consist of three closely spaced energy levels even in the absence of an external field, except in highly symmetric molecules [19] (in spherical and cubical symmetry all spin-spin interactions cancel so that $D = E = 0$). Zero field splitting is quantified by the first order zero field splitting constant D (about 0.1 cm^{-1}) and the second order constant E (about 0.01 cm^{-1}). These constants are indicated in Fig. 9.15 for one particular orientation of the spin system vs. the external magnetic field. The single resonance at B_0 (Fig. 9.14) is replaced by two $\Delta m_s = 1$ resonances B_a and B_b which strongly depend on the orientation of the molecular axes relative to the external field. A full analysis of zero field parameters can hence only be made in samples of known molecular orientation. The half-field resonance is less orientation-dependent and can easily be recognised even in amorphous, rigid solutions. From the exact position of this resonance (B_{min}) a mean splitting parameter D^* can be deduced [Eq. (9.16)] which gives an approximate measure of the strength of interaction and thus of the mean distance of the two electron spins forming the triplet system (D^* is roughly inversely proportional to the mean distance of the two unpaired electrons forming the triplet state).

$$D^* = \sqrt{D^2 + 3E^2} = \sqrt{\frac{3\,(h\nu)^2}{4} - 3\,(g_e\mu_B B_{min})^2}. \tag{9.16}$$

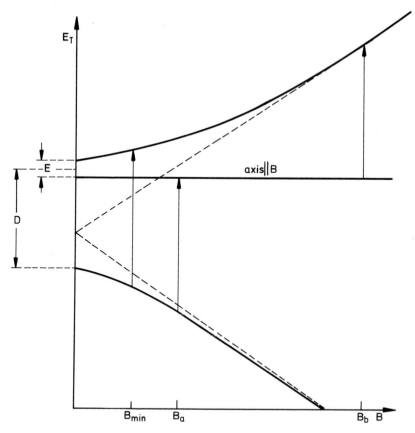

Fig. 9.15. Zero field splitting of a triplet of relatively low summytry (one particular orientation relative to the external field). The length of the arrows corresponds to MW photon energy used; dashed lines correspond to $M_s \pm 1$ in Fig. 9.14

The total zero field splitting is approximately given by D (or D^*). Strong magnetic fields decouple the triplet state so that the two spins are quantised separately, as shown in Fig. 9.14 (m_s is a good quantum number in this high field case, but not so in the zero field region).

In symmetrical molecules, the magnetic moment of each of the three (or two if $E=0$) zero field levels of the triplet corresponds to one main axis of the molecule or part of macromolecule. In molecules of medium symmetry ($E>0$) up to six $m_s=1$ peaks can be observed in the ESR spectrum. If the triplets are oriented in a single crystal, they can collectively be rotated with regard to the external magnetic field so that the resonances are observed one after the other.

This experiment has been performed e.g. using partly polymerised TSHD single crystals (see also Sect. 6.6) [20, 21]. The triplet, in this exceptional case, is an electronic ground state related to the growing chain ends of the TSHD macromolecules:

TSHD (Carbene-type chain end)

$R = -CH_2-O-SO_2-\!\!\left\langle\bigcirc\right\rangle\!\!-CH_3$

Structures of this type are called carbenes which in principle can also form singlets ($\uparrow\downarrow$). The triplet nature ($\uparrow\uparrow$) of the growing chains has been evidenced by ESR absorption, since the singlet carbene does not give any ESR signal. The large zero field splitting constant of polymerising TSHD

$$D = 0.27 \text{ cm}^{-1}$$

points to close neighbourhood of the spins in the reactive center, causing strong magnetic interaction.

In the low temperature photopolymerisation of TSHD crystals other ground state triplets ("diradicals") have been detected in addition to the triplet carbenes and even dicarbenes forming quintet states were identified by ESR spectroscopy [33-35]. These studies which have been reviewed recently by Sixl [36] give a very detailed picture on the primary processes of the solid state polymerisation in diacetylene single crystals.

Most triplets observed by ESR are excited electronic states (Sect. 4.5.1). The study of polymer triplets, e.g. in aromatic and ketonic polymers should be much more used as a method complimentary to phosphorescence spectroscopy [22]. The most promising field of polymer applications for ESR spectroscopy of these "optical" triplets seems to be the elucidation of trapping centres which have been identified by emission spectroscopy. Triplet excitons may also be investigated by ESR although their concentration is in general smaller so that sensitivity problems are likely to occur.

It should be added that zero field splitting can be investigated at low temperature without external fields and that transitions between the sublevels can be induced by microwaves and observed using phosphorescence as a probe (double resonance techniques) [23].

9.8 Spin-Labels

Diamagnetic polymers can be converted into ESR-active derivatives using stable radicals, e.g. nitroxides:

These radicals can either be attached by covalent bonds to the polymer (spin labels) using reactive substituents (R) or dissolved physically (spin probes) [12, 24].

133

D. Spin-Resonance Spectroscopy

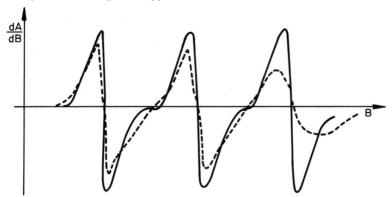

Fig. 9.16. ESR of a typical nitroxide spin label in fluid (——) and viscous (---) solution; the splitting is due to ^{14}N ($a_N \approx 15$ G)

Nitroxides show a well resolved ^{14}N-split (Fig. 9.16) hyperfine spectrum in solvents of low viscosity. In viscous solution, anisotropic broadening first decreases the intensity of the high-field peak (Fig. 9.16).

Broadening is observed when the spin label needs more than about 10 ps for one rotation. Hindered rotation may be due to firm linking with a macromolecule. The spectrum therefore indicates the frequency of polymer movements. The total time regime available for the measurement amounts to

$$\tau_c \text{ (spin-label)} \approx 10 \text{ ps to } 1 \text{ } \mu s$$

and overlaps with the time range of fluorescence polarisation (fluorescence label) technique. Average rotation times (τ_c) are calculated from peak broadening [12, 24]. In order to exclude electron spin-spin interaction, a dilution of 1:200 to 1:600 (mol label/mol basic unit) is generally attempted. It is often possible to label the end group(s) of macromolecules specifically (PEO, polyester, polyamides, biopolymers). Alternatively, substitution is performed along the chain.

Examples (tert. butyl = +):

PS

mole ratio label to basic unit: 1/160

134

Typically, the average time needed for rotations τ_c, as a measure of inverse chain or end group flexibility, increases with $\bar{M}_n$, as in the case of labelled PS in toluene.

A limiting value $\tau_c = 400$ ps is reached at about $\bar{M}_n = 10^5$. This effect is due to the superposition of movements of the polymer coil as a whole (τ_{tot}) and "local modes" (τ_{lm}) which are due to movements of side groups and segments [Eq. (9.17)]

$$\frac{1}{\tau_c} = \frac{1}{\tau_{tot}} + \frac{1}{\tau_{lm}}, \tag{9.17}$$

where τ_{tot} can be estimated from viscosity data [24, 25] according to Eq. (9.18)

$$\tau_{tot} = A M [\eta] \, \eta_{Solv} / RT \tag{9.18}$$

A (free draining coil) $= 0.61$, A (non draining coil) $= 0.42$, $[\eta] =$ limiting viscosity number, $\eta_{Solv} =$ viscosity of solvent.

The temperature dependence of τ_{tot} reflects the activation of η_{Solv} (9 kJ mol^{-1} in toluene); τ_{lm}, on the other hand, is found to have an activation energy of 18 kJ mol^{-1} (PS), similar to local mode activation in (pCLPS) – and pFPS measured using dielectric relaxation techniques.

The spin label technique has recently been reviewed by Törmälä et al.[37] and by Cameron[38].

9.9 Résumé

ESR, except for specially prepared polymers, gives no direct information about the chemical nature of the intact polymers (Table 9.3), since these in general are diamagnetic; it does, however, give precise information about the radicals formed by radiation, mechanical stress, etc. The spin density is closely related to the electronic structure of doublet and triplet states which can be studied by ESR. High resolution allows the spin density to be measured using Mc Connell's equations analogous to Eq. (9.4).

The chain movements are studied using the spin label technique which in this application is complementary to several other relaxation methods giving similar information. In exciton studies, triplet excitons and their traps in aromatic polymers are likely objects of further research. Polymeric CT complexes[26] showing at least partial elec-

Table 9.3. Informations obtained by ESR relating to

Structure		Dynamics	
of polymeric systems			
Chemical structure	+	Movements of the chain, segments and side groups	+
Tacticity	−		
Conformation	−	Phonons	−
Crystallinity	−	Excitons	(+)
Electronic structure	+	Complex formation and related phenomena	+

tron transfer from donor (e.g. PVCA) to acceptor molecules can be studied using ESR. The paramagnetic species formed can be regarded as radical cations and anions which are closely related to charge carries in polymers (hopping model)[26]. Highly conjugated polymers mostly contain appreciable concentrations of unpaired electrons and have been studied by ESR in order to explain the catalytic and conductive properties of these polymers.

Electron Spin Research is unique for studying the primary processes of mechanical degradation and similar degradation processes following a radicalic pathway (e.g. photolysis). If highly reactive primary products are to be studied, low temperature or high pressure[27] is needed in order to moderate possible secondary reactions.

ESR spectroscopy is a highly specific and if relaxation is neither too slow nor too fast, also an extremely sensitive method. It is not, however, an analytical routine method in polymer research.

References

1. Rånby, B., Rabek, J.F.: ESR Spectroscopy in Polymer Research. Berlin, Heidelberg, New York: Springer 1977; this book contains 2,519 references, up to 1975
2. Swarz, H.M., Bolton, J.R., Borg, D.C.: Biological Applications of Electron Spin Resonance. New York: Wiley 1972
3. McConnell, H.M., Chesnut, D.B.: J. chem. Phys. 28, 107 (1958)
4. Rembaum, A., Moacanin, J., Haak, R.: J. Macromol. Chem. 1, 657 (1966)
5. Ayscough, P.B.: Electron Spin Resonance in Chemistry. London: Methuen 1967
6. Atherton, N.M.: Electron Spin Resonance, Theory, and Application. New York: Halsted 1973
7. Geschwind, S.: Electron Paramagnetic Resonance. New York: Plenum 1971
8. McMillan, J.A.: Electron Paramagnetism. New York: Reinhold 1968
9. Scheffler, L., Stegmann, H.B.: Electron Spin Resonance. Berlin, Heidelberg, New York: Springer 1970
10. Fischer, H.: In: Hummel, D.O. (Ed.): Polymer Spectroscopy. Weinheim: Verlag Chemie 1974, p. 289
11. Kausch, H.H.: Polymer Fracture. Berlin, Heidelberg, New York: Springer 1978
12. Törmälä, P., Lindberg, J.J.: In: Iving, K.J. (Ed.): Structural Studies of Macromolecules by Spectroscopic Methods. London: Wiley 1976, p. 255
13. Keighley, J.H.: In: Jones, D.W. (Ed.): Introduction to the Spectroscopy of Biological Polymers. New York: Academic Press 1976, p. 221
14. Fischer, H.: Advances Pol. Sci. Vol. 5, 463 (1967)
15. Yoshida, H., Hayashi, K.: Adv. Pol. Sci. Vol. 6, 401 (1969)
16. Kausch, H.H., Becht, J.: Kolloid-Z. u. Z., Polymere 250, 1048 (1972)
17. Carstensen, P., Rånby, P.: In: Radiation Research. Amsterdam: North Holland Publ. 1967, p. 297
18. Sohma, J., Sakagudri, M.: Adv. Pol. Sci. Vol. 20, 1 (1976)
19. McGlynn, S.D., Azumi, T., Kinoshita, M.: Molecular Spectroscopy of the Triplet State. Englewoods Cliff: Prentice Hall 1969
20. Stevens, G.C., Bloor, D.: Chem. Phys. Lett. 40, 37 (1976)
21. Eichele, H., Schwoerer, M., Huber, R., Bloor, D.: Chem. Phys. Lett. 42, 342 (1976)
22. Rippen, G.: Optische und ESR-spektroskopische Untersuchungen von molekularen Komplexen im Triplettzustand. Göttingen: Thesis 1976
23. El-Sayed, M.A.: In: Adv. Photochem. Vol. 9, 311 (1974), New York: Wiley, and in: Lim, E.C. (Ed.): Excited States, Vol. 1, 35 (1974), New York: Academic Press
24. Bullock, T.A., Cameron, G.G.: In: Ivin, K.J. (Ed.): Structural Studies of Macromolecules by Spectroscopic Methods. London: Wiley 1976, p. 273

25. Zimm, B.H.: J. Chem. Phys. *24*, 269 (1956)
26. Mort, J., Pfister, G. (Eds.): Electronic Properties of Polymers. New York: Wiley 1982
27. Szöcz, F.: J. Appl. Pol. Sci. *27*, 1865 (1982)
28. Rånby, B.: In: Bayer, F., Keinath, S.E. (Eds.): Molecular Motion in Polymers by ESR. New York: Harwood acad. publ. 1980
29. Bresler, S.E., Kazbekov, E.N., Saninskii, E.M.: Vysokomol. Soedin. *1*, 132 (1959)
30. Bamford, C.H., Jenkins, A.D., Symons, M.C.R., Townsend, M.G.: J. Pol. Sci. *34*, 181 (1959)
31. Ingram, D.J.E., Symons, M.C.R., Townsend, M.G.: Trans. Faraday Soc. *54*, 409 (1958)
32. Canbäck, G., Rånby, B.: Macromolecules *10*, 797 (1977)
33. Huber, R., Schwoerer, M.: Chem. Phys. Lett. *72*, 10 (1980)
34. Bubeck, C., Sixl, H., Neumann, W.: Chem. Phys. *48*, 269 (1980)
35. Neumann, W., Sixl, H.: Chem. Phys. *58*, 303 (1981)
36. Sixl, H.: Spectroscopy of the Intermediate States of Solid State Polymerisation in Diacetylene Crystals, submitted to Advances in Polymer Science
37. Törmälä, P., Weber, G., Lindberg, J.J.: In: Boyer, R.F., Keinath, S.E. (Eds.): Molecular Motion in Polymers by ESR. New York: Harwood acad. publ. 1980
38. Cameron, G.G.: Pure Appl. Chem. *54*, 483 (1982)

10 Nuclear Magnetic Resonance (NMR) Spectroscopy of Polymers

10.1 The Origine of NMR Spectra

Compared to ESR Spectroscopy, the interaction between the magnetic particles and the external magnetic field is much weaker in NMR. The resonance frequency is therefore much smaller and can be found in the range of radio frequencies (corresponding to wavelengths of 1 to 10 m) [1, 2]. As discussed in Chap. 8, the main reason for this behaviour is the weakness of nuclear magnetic moments. Table 10.1 shows a list of nuclei frequently encountered in polymers together with their natural abundance and relative sensitivity of detection. The nuclear spin moment is given in Table 10.1 in units of $h/2\pi$.

The spin of protons and neutrons (1/2) adds up to the total nuclear spin (0, 1/2, 3/2...) which, if non-zero, causes the magnetic moment of the isotope considered. The usefulness of NMR as a tool in organic polymer research rests in the magnetic moment of several important nuclei (^1H, ^{13}C, ^{14}N, ^{19}F...) and in the absence of such a moment in the main isotope of carbon, ^{12}C (if ^{12}C was a magnetic isotope, the NMR spectra of most organic compounds were excessively complicated). In the following sections we shall first consider ^1H-NMR spectra which until recently dominated polymer spectroscopy due to the relatively high sensitivity of ^1H [3]. Only ^{19}F shows a comparable sensitivity and therefore can be studied using the classic (continuous mode) instrumentation discussed in Sect. 10.2.

The proton spin moment of 1/2 ($m_s = \pm 1/2$) causes a splitting into two energy levels, in analogy to unpaired electrons in ESR (10.1):

$$\Delta E = h\nu = g_N \mu_N B \tag{10.1}$$

g_N: nuclear g-factor ($g_p = 5.5857$ in case of proton)
μ_N: nuclear magneton (see Chap. 8).

In Table 10.2, experimental resonance frequencies, the corresponding wavelengths and magnetic flux densities are given for the free proton, neglecting any screening effects.

D. Spin-Resonance Spectroscopy

Table 10.1. Nuclear spin of important isotopes

Element	Main isotope nuclear spin = 0	Isotopes nuclear spin > 0 (natural abundance percent)	Approximative sensitivity relative to ^{1}H
H		^{1}H: 1/2 (99.98)	1
		^{2}H = D: 1 (1.5×10^{-2})	10^{-2}
C	^{12}C	^{13}C: 1/2 (1.11)	1.6×10^{-2}
N		^{14}N: 1 (99.64)	10^{-3}
		^{15}N: 1/2 (0.36)	10^{-3}
O	^{16}O	^{17}O: 5/2 (3.7×10^{-2})	3×10^{-2}
F		^{19}F: 1/2 (100)	0.83
Si	^{28}Si	^{29}Si: 1/2 (4.7)	8×10^{-3}
P		^{31}P: 1/2 (100)	0.07
S	^{32}S	^{33}S: 3/2 (0.74)	2×10^{-3}
Cl		^{35}Cl: 3/2 (75.4)	5×10^{-3}
		^{37}Cl: 3/2 (24.6)	3×10^{-3}

Table 10.2. NMR resonance conditions for protons

Resonance frequency v_0	Wavelength	Magnetic flux density ($1T = 10^4$ G)
60 MHz = 6×10^7 s^{-1}	5 m	1.4 T
100 MHz = 10^8 s^{-1}	3 m	2.3 T
300 MHz = 3×10^8 s^{-1}	1 m	7.0 T

This (free) proton resonance frequency v_0 is most easily calculated according to (10.2).

$$v_0 \text{ (MHz)} = 42.577 \times B\,(T)\,. \tag{10.2}$$

Using strong magnetic fields turns out to be advantageous for two reasons:
- Increasing ΔE improves the Boltzmann distribution N_s/N_0 between upper and lower state [Eq. (9.5)]. ΔE at 100 MHz (0.4 mJ mol^{-1}) corresponds to only $1.6 \times 10^{-5}\ kT$ at room temperature so that N_s/N_0 is very near to 1 anyway. The increased N_s/N_0 enhances the sensitivity of NMR measurements.
- The resolution of spectra is improved by high fields, too, since the chemical shift is proportional to the strength of the applied field, but nuclear spin-spin coupling is not.

For chemical applications of NMR, the following effects are of prime importance:
(1) Magnetic Dipole-Interaction (anisotropic shielding)
(2) Chemical Shift
(3) Nuclear Spin-Spin Coupling

Magnetic Dipole-Interaction. Each magnetic nucleus belonging to solvent molecules or to dissolved macromolecules, is a micro magnet which slightly modifies the magnetic field experienced by neighbouring protons so that the resonance field is somewhat different for each proton even if they are chemically equivalent. It follows a peak broadening which in the solid state prevents high resolution spectra of polymers, unless a special geometrical arrangement (magic angle) is combined with NMR pulse techniques (see Sect. 10.8.2).

Classical solid polymer spectra are "broad-line" spectra giving information about movements, but virtually zero information on chemical structure (Sect. 10.4).

Due to the rapid rotation of flexible macromolecules and their segments in solution, the above interactions are averaged to zero, so that in low viscosity solutions high resolution spectra can be observed. However, in rigid polymers and viscous solvents, restrictions of molecular mobility are frequently observed so that the spectral resolution is poor in these cases.

In practice, proton-free solvents of low viscosity are used (CCl_4, $CDCl_3$) whenever possible. In the case of rigid polymers, the temperature is increased in order to facilitate rapid molecular movements.

Chemical Shift. The interaction of a magnetic field with matter is not restricted to magnetic moments already present in the sample, but also induces magnetic moments. This diamagnetism causes the magnetic shielding of protons and other nuclei to be studied[4]. If B_0 is the resonance magnetic flux density of the free proton according to Eq. (10.1) and (10.2), it is modified by diamagnetic shielding according to (10.3)

$$B_i = B_0 (1 - \sigma_i) , \tag{10.3}$$

where σ_i, the shielding constant is composed of a diamagnetic positive term σ_d and a paramagnetic, negative term σ_p:

$$\sigma_i = \sigma_d + \sigma_p . \tag{10.4}$$

In the case of H, the diamagnetic term dominates, so that σ_H is positive and, hence, $B_i < B_0$, i.e. the magnetic flux density is smaller for protons in molecules compared to "nacked" protons. A given proton, corresponding to resonance condition in Eq. (10.1), is now off-resonance due to the lower effective magnetic flux density B_i. The external field has to be increased in order to achieve resonance. It is chemically of outmost importance that σ_i strongly differs for differently bound protons; it is lowest for "acid" H-atoms.

This is shown by the well-known text book example of (medium resolution) ethanol ^1H-NMR absorption; three peaks whose relative areas are 1:2:3, in the order of increasing B. Obviously, the three peaks correspond to O-H, CH_2, and CH_3 protons, the methyl group experiencing the strongest shielding, the "acid" hydroxyl proton the smallest one. In NMR-spectroscopical practice, chemical shifts (ϑ_i) are given relative to an internal standard substance

$$\vartheta_i = (\delta_{\mathrm{ref}} - \sigma_i) \times 10^6 \ (\mathrm{ppm}) . \tag{10.5}$$

The sign in Eq. (10.5) has been recommended by IUPAC; ϑ_i is accordingly positive for all protons which are *less* shielded than the standard. This is true for nearly all types of H if the strongly shielded TMS is used as standard [ϑ_p (free proton) $= 31$ ppm].

Splitting   NMR

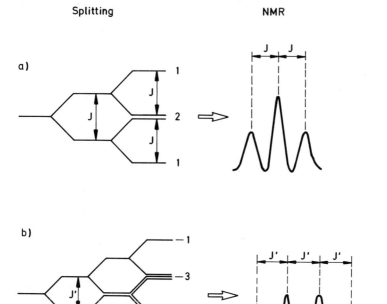

Fig. 10.1 a, b. Spin-spin coupling of the protons of C_2H_5–OH (schematic drawing); **a** CH_3 group, coupling constant J; **b** CH_2 group, coupling constant J' (no O–H splitting; rapid proton exchange limit)

The internal standard compensates shielding effects of the solvent and the sample tube wall. Experimentally, ϑ_i is determined according to Eq. (10.6) [5].

$$\vartheta_i = \frac{B_{ref} - B_i}{B_{ref}} \times 10^6.$$ (10.6)

A large chemical shift, e.g. for R-COOH, is in the order of 10 to 15 ppm, relative to TMS; a shift of 1 ppm (at $v = 100$ MHz) corresponds to $\Delta v = 100$ Hz or $\Delta B = 23$ mG.

Nuclear Spin-Spin Coupling. Spin-spin coupling allows the relative position of protons to be determined and can be understood only on the basis of quantum mechanics [magnetic dipole interaction (1) and diamagnetic shielding (2) can be understood using classical models]. In order to illustrate the influence of spin-spin coupling on NMR spectra, the well-known high resolution spectrum of C_2H_5OH should be considered: the central CH_2 band splits into four peaks (1:3:3:1) and the high field CH_3 band splits into three peaks (1:2:1) if the resolution is high enough. According to the rules presented in Chap. 9 and taking into account that the spin quantum number of the proton is $m_s = \pm 1/2$, the experimentally observed splittings can easily be explained, provided there is a cou-

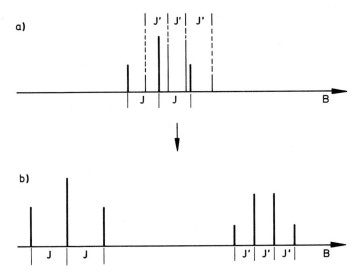

Fig. 10.2. Improvement of spectral resolution from low (**a**) to high (**b**) magnetic flux density (B). The spin coupling constants J and J' are not influenced by B, the evident simplification results from changing chemical shifts which are proportional to B

pling mechanism which connects the neighbouring non equivalent protons (Fig. 10.1). The highest sensitivity may also reveal spin-spin coupling between more distant protons.

Figure 10.1 shows that
- The magnetic moments of the CH_3 group couple with those of the two methylene protons
- The protons of the methylene groups couple with all three protons of the methyl group
- No splitting occurs between the equivalent protons of the same group
- The interaction occurs through the distance of three bonds.

The question therefore arises how the magnetic moments "feel" each other over the distance. Evidently, the electronic system of the molecule acts as a mediator. The magnitude of the splitting (J) is decisively influenced by the distance and relative orientation of the protons, but is independent of the strength of the external magnetic field. This behaviour is analogous to, e.g., zero field splitting in triplet states and electron-nuclear spin coupling observed in ESR.

This independence of magnetic field strength (or flux density B) is the main reason of the spectral simplification achieved by increasing B, as shown schematically in Fig. 10.2.

The art of interpreting NMR spectra consists in detecting related peaks, finding the type of bonding (from chemical shift) and identifying the proton positions relative to neighbouring protons (from spin-spin coupling). In technical polymers, resins, etc. which often contain plastisisers and other additives, the ^{1}H-NMR spectra are in general quite complicated. The analytical-chemical usefulness of these spectra, at least in routine work is therefore limited.

D. Spin-Resonance Spectroscopy

The order of magnitude of the spin-spin coupling constant ist

$$J \approx 1 \text{ mG} \hat{\approx} 10 \text{ Hz} .$$

The homogeneity of the magnetic field, therefore, has to be of the order of $1:10^7$ to $1:10^8$ to allow the resolution (J).

We now return to the C_2H_5OH spectrum in which splitting of the hydroxyl proton is not observed, at least not in only moderately pure samples. The behaviour is to be attributed to proton exchange processes, which are due to traces of acids or bases, e.g. (10.7):

$$C_2H_5OH + H^+ \rightarrow C_2H_5OH_2^+$$

$$C_2H_5OH_2^+ \rightarrow C_2H_5OH + H^+ . \tag{10.7}$$

These processes, which are closely connected with Grotthus-(proton hopping) conductivity of protonic substances, decrease the average time spent by a proton at the OH group to such an extent that spin-spin coupling cannot occur. Any rapid exchange process has the same consequence if

$$\tau \leqq \frac{1}{\Delta \nu (=J)} \tag{10.8}$$

τ: average resdence time of H at the group causing the signal (s)
J: spin-spin coupling constant (Hz).

If $J = 10$ Hz, the proton has to spent at least $\tau = 0.1$ s at a particular group (e.g. OH) in order to produce a measurable peak splitting. If the exchange is faster, i.e. τ smaller, this splitting vanishes. In high-purity ethanol, the OH signal splits into a 1:2:1 "triplet" as expected for coupling with the central CH_2 groups, ^{16}O being magnetically inert (Table 10.1). In this case, also the peaks of the CH_2 group double by interaction with the single proton at OH. Addition of a few percent of very pure water does not alter this well resolved spectrum, addition of more than 25 percent H_2O, however, causes the appearance of a peak averaged between OH (C_2H_5OH) and OH (H_2O) indicating an averaged chemical shift. This averaging can be used in order to study slow proton exchange processes, rotations etc.[6]. Spin-spin coupling and chemical shift averaging, therefore, are useful kinetic probes for slow processes.

Often, however, it is desirable to make spectra simpler by removing spin-spin coupling. This can be achieved experimentally by irradiating the sample using the characteristic frequency of the proton to be decoupled, thus destroying the interaction with its neighbours.

A further requirement of high resolution is narrow line width. Saturation, therefore, has to be avoided; the Boltzmann distribution between lower and upper states is very unfavourable anyway; equal distribution removes any measurable NMR absorption, since the RF field induces exactly as many "down" as "up"-transitions in this case. The long spin-lattice relaxation time of protons

$$T_1 \approx 1 \text{ s}$$

favours saturation effects; T_1 can be decreased by paramagnetic impurities such as molecular oxygen.

142

Chemical reactions may disturb the Boltzmann distribution in such a way that the emission of RF quanta (CIDNP) can be observed[30].

10.2 Experimental

The essential parts of a high resolution NMR spectrometer are shown in Fig. 10.3:

The *radiation source* is an emitter coil envelopping the sample tube. The emitter coil is connected with the RF generator and can be simultaneously used as detector if a bridge circuit is used, as indicated in Fig. 10.3. The magnetic field produced by the coil is perpendicular to the external one.

The *magnet* has to produce magnetic flux densities of the order of

$$B \approx 1 \text{ to } 7 \; T \; .$$

Permanent magnets are relatively simple, but very high flux densities cannot be achieved in this way and the temperature has to be accurately controlled. Homogenous standard electromagnets can be used up to about 2.5 T, whereas supraconducting coils are suitable as magnets for very high magnetic fields (11.7 T for ^{1}H-NMR at 500 MHz)[46] . These coils have to be cooled by means of liquid He.

Modulation. The NMR spectrum can be produced by varying the magnetic field strength using an additional coild (field sweep) or by varying the frequency of the emitter using a.f. modulation (frequency sweep).

The *sample* is contained in a tube spinning around the long axis in order to improve the effective homogeneity of the magnetic field.

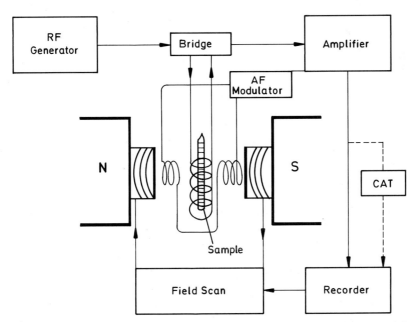

Fig. 10.3. Block diagram of a (continuous mode) NMR spectrometer, as used for ^{1}H- and ^{19}F-NMR spectroscopy

If the *detector* is not identical with the emitter coil, a detector coil can be inserted perpendicular to the external and the emitter fields. This coil measures the magnetisation of the sample by detecting the induced a.f. tension.

The NMR *signal* can be recorded after amplification provided the S/N ratio is sufficient. Otherwise, signal averaging has to be applied. This procedure requires extreme constancy of the experimental conditions (feed back techniques) ensuring that the signal appears at exactly the same resonance position in repetitive scans. This improvement is especially important in polymer spectroscopy, since concentration should be small in order to keep the viscosity low. High viscosity broadens the absorptions peaks and thus diminishes the spectral resolution. For the same reason, temperature has to be increased in many cases.

Fourier-Transform NMR-spectroscopy will be discussed together with ^{13}C-NMR in Sect. 10.8.

10.3 High-Resolution ^{1}H-NMR of Polymers

10.3.1 Applications

High-resolution ^{1}H-NMR spectra of polymers[3, 47)] have to be recorded in solutions of low viscosity. In Fig. 10.4, the order of magnitude of the different line-broadening and shifting mechanisms is indicated. If it is possible to achieve the necessary small line width, the following informations can be gained from the ^{1}H-NMR spectrum:

– Identification of the polymer by comparison with reference spectra recorded under the same experimental conditions – above all, the same solvent and the same resonance frequency has to be used. Collections of reference spectra are available for $v_0 = 60$ and 100 MHz[9)]. The approach discussed here is similar to the one used in em-

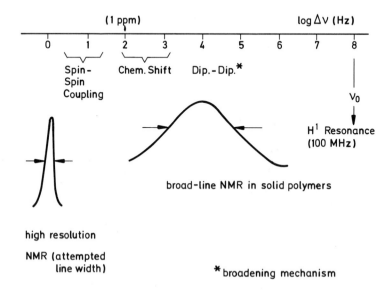

Fig. 10.4. Schematic view of peak broadening and shifting due to different mechanisms at a frequency scale

pirical mIR spectroscopy and may be successful in identification of well-defined pure polymers.
- Chemical shift, spin-spin coupling and number of protons. This evaluation corresponds exactly to the one applied to unknown low molar mass compounds. The spectra may be much more complicated due to higher order spin-spin coupling especially if J is nearly equal to the difference of the chemical shifts of the coupling protons [2].
- Configuration (tacticity-) analysis; this application of polymer NMR spectroscopy gives results not available by any other method [2, 3, 10, 11] except by other, more advanced NMR techniques (Sect. 10.8). Examples for this most important application are given in the following section.
- Dynamical data are obtained using line broadening, chemical shift averaging etc.

10.3.2 Tacticity Analysis

Tacticity analysis by ^{1}H-NMR is discussed in the following, starting with the now classical example of PMMA, the first polymer which was successfully examined by means of this method for its tacticity in 1960 [12-14]. This polymer seems to be extremely suitable for this work since measurable ^{1}H spin-spin coupling may be ruled out because of the chemical structure of the basic unit;

PMMA

All non-equivalent protons are separated from each other by more than three bonds. Hence, any ^{1}H-NMR splittings which may be observed in PMMA can only be due to the inequivalency of protons induced by the steric arrangement (conformation) of the macromolecule which is controlled by its configuration or tacticity.

Actually, some PMMA samples – later on to be identified as highly syndiotactic ones – show a very simple ^{1}H-NMR spectrum [13] (Fig. 10.5). In some PMMA samples (e.g. those polymerised with a Grignard's reagent) two essential differences can be observed: the C-CH$_3$ peak is shifted from $\vartheta = 0.91$ to 1.20 ppm and the CH$_2$ signal at 1.82 ppm is split into a quartet. The O-CH$_3$ group, which is situated at the periphery of the macromolecule is unaffected ($\vartheta = 3.59$ ppm). The splitting of the CH$_2$ signal in the poly-

145

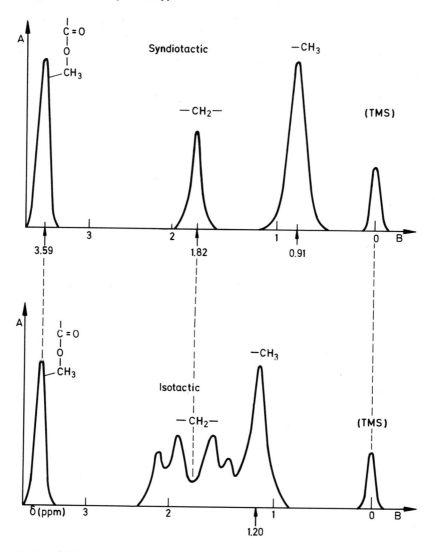

Fig. 10.5. ^{1}H-NMR absorption of PMMA, schematic after U. Johnsen [10, 13)]

mer labelled "isotactic" in Fig. 10.5 can only be due to some inequivalence of the two protons induced by the stereochemistry of the polymer leading to geminal coupling. Methyl groups, on the other hand, rotate very fast so that an inhomogenous "micro environment" – as indicated by CH_2 splitting, cannot induce any measurable C-CH_3 splitting.

In order to explain this important experiment, we first have to recall the meaning of the term "tacticity" which has been introduced by Natta in 1955 [15)] (see also Sect. 7.4.3).

Natta called those molecules of vinyl polymers isotactic, which show the same configuration at all pseudo-asymmetric C*-atoms (ddd ... or lll ...):

Extended chain projection of an isotactic vinyl polymer.

Molecules with alternating d and l configuration (dldl...) have been called syndiotactic.

Extended chain projection of a syndiotactic vinyl polymer.

Any less regular structure originally was termed atactic, whereas more recently the term atactic is more frequently used for polymers showing random sequences. A better characterisation is obtained if the relative number of d and l links is measured and if additionally longer blocks of isotactic or syndiotactic configuration, if present, are identified. Studies of this type, made possibly by NMR, reveal the "micro-tacticity" [10] of polymers. Due to their regular micro structure polymers of well-defined tacticity can form macroscopic crystal lattices so that they may also be studied by means of diffraction methods. Configurationally less ordered polymers, on the other hand, do not crystallise so that diffraction methods cannot be applied; high resolution NMR is the most direct approach in structural studies of this type.

In order to understand the different spectral features of PMMA samples supposed to show different tacticity, we first consider the extended zig-zag chain of isotactic PMMA:

Isotactic PMMA, extended chain

It is easily noted that in this particular conformation of isotactic PMMA the two protons of CH_2 are not equivalent with regard to their environment; the protons labelled "a" are nearer to the ester groups, those labeled "b" nearer to C-CH_3. This inequi-

147

valence of the methylene protons persists on the average also in all other conformations of the isotactic configuration, which can be formed by rotations about the C-C bonds of the main chain.

A different situation is encountered in the syndiotactic polymer, as can be seen from the extended zig-zag conformation:

Syndiotactic PMMA

Here, the "right-hand"- and "left hand" neighbours of each CH_2 proton are different – either $C-CH_3$ or $COOCH_3$; therefore, they "feel" on the average the same environment. Splitting of the CH_2 NMR peak which is due to spin-spin coupling between inequivalent protons is excluded in this case and hence the interpretation given in Fig. 10.5 is correct and unambiguous. Furthermore, the difference in the chemical shift of $C-CH_3$ observed in PMMA samples of different tacticity can be used in order to quantify the relative abundance of the three possible "triades". This can be explained on the basis of the above zig-zag conformations, focussing our attention on the central CH_3 groups (encircled): in the isotactic triade, two other $C-CH_3$ groups are the nearest neighbours, whereas in the syndiotactic triade the neighbours are ester groups. The third possible arrangement is that where two different configurations join (heterotactic triade). In this case, the central $C-CH_3$ group is next to one $C-CH_3$ and one ester group. Whatever the reason for the different diamagnetic shielding ($\vartheta_{syndio} = 0.91$ ppm; $\vartheta_{iso} = 1.20$ ppm) may be, the chemical shift of the $C-CH_3$ of heterotactic triades is expected to lie between the two extreme values quoted.

Actually, in configurationally inhomogenous ("atactic") PMMA, a third $C-CH_3$ peak is found at $\vartheta_{hetero} = 1.04$ ppm.

The relative proportion of the three triades, which is calculated from the areas under the CH_3-iso, syndio and hetero peaks, can be used as measure of "micro tacticity" for those polymers which cannot be classified as 100 percent isotatic or 100 percent syndiotactic, i.e. for the vast majority of polymer samples, showing pseudo-asymmetric centres.

In order to quantify the microtacticity of linear macromolecules, two quantities have to be known [3]:

– The relative amount of meso (m) and racemic (r) placements between neighbouring basic units of the polymer. These formations are called "dyads". Correspondingly, three consequentive basic units may form either an isotatic (i = mm), heterotactic (h = mr, rm) or syndiotactic (s = rr) "triad".

– The average lengths of isotactic and syndiotactic sequences $[l(i), l(s)]$ and their distribution functions.

The symbols used for dyads are $\overline{m}$ and $\overline{r}{\vert}$.

Dyads have either meso (dd, ll) or racemic (dl, ld) configuration. Triads can be symbolised as follows:

$$\overline{m} \qquad \overline{r} \qquad \overline{mr} \qquad \overline{rm}$$

$$i = mm \qquad\qquad s = rr \qquad\qquad mr \quad (h \text{ or } \overset{\frown}{mr}) \qquad rm .$$

Experimentally, only i, s, and h can be distinguished, whereas mr is equivalent to rm. In the example of PMMA, the intensity of the C-CH$_3$ signal at $\vartheta = 1.04$ ppm is proportional to the concentration of the heterotactic triads:

$$I_{0.91} \sim P(rr)$$

$$I_{1.04} \sim P(mr) + P(rm) \qquad\qquad (10.9)$$

$$= P(\overset{\frown}{mr})$$

$$I_{1.20} \sim P(mm) .$$

If the intensity of the individual peaks is normalised to the total intensity of the C-CH$_3$ signal, the relative concentrations or probabilities (P) of the different triads are obtained:

$$I_{rel} = \frac{I_{0.91}}{\Sigma I} = P(rr) \text{ etc.} \qquad\qquad (10.10)$$

Applying Eq. (10.10) requires that the molar mass ($\bar{M}_n$) of the polymer is relatively large so that end group effects can be neglected. The concentration of the triads can be deduced from the NMR spectrum directly according to Eq. (10.10), provided its resolution is sufficient. The relative concentration of dyads, P (m) and P (r) can be derived according to the following considerations [10]:

Since a meso placement can only follow up on a meso or racemic placement,

$$P(mm) + P(rm) = P(m) . \qquad\qquad (10.11)$$

It is equally true that a meso placement can only precede a meso or racemic placement:

$$P(mm) + P(mr) = P(m) . \qquad\qquad (10.12)$$

Hence, we find from Eqs. (10.11) and (10.12) that

$$P(rm) = P(mr) = 1/2(P(rm) + P(mr)) , \qquad\qquad (10.13)$$

$$P(m) = P(mm) + 1/2(P(rm) + P(mr)) . \qquad\qquad (10.14)$$

The relative concentration of racemic dyads is given by $P(r) = 1 - P(m)$.

Including Eq. (10.10), we now have derived the relative concentrations of meso and racemic placements or dyads from experimental information. These dyads, however,

D. Spin-Resonance Spectroscopy

Table 10.3. Intensities of C-CH$_3$ peaks of PMMA, 220 MHz in Chlorobenzene [16)]

	i (rr)	h (mr)+(rm)	s (mm)	Sum
Peak (cm)	5.3	3.2	0.5	9.0
Relative intensity ($= P$)	0.59	0.36	0.05	1

can form many short sequences or a few long ones. It can immediately be seen that in this latter case the concentrations of heterotactic triads has to be small whereas in the former case of short sequences P (h) has to be large. Since each isotactic sequence, formed by meso dyads is terminated by a racemic dyad

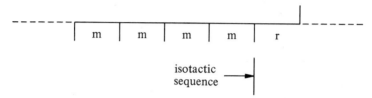

is the relative concentration of mr triads equal to the relative concentration of isotactic sequences. The average length $l(i)$ is given by

$$l(i) = \frac{P(m)}{P(mr)}$$ (10.15)

and

$$l(s) = \frac{P(r)}{P(mr)}.$$ (10.16)

As an example, a commercial PMMA sample will be analysed in the following (Table 10.3). According to Eq. (10.14) we obtain

$$P(m) = 0.05 + 0.18 = 0.23$$
$$P(r) = 0.59 + 0.18 = 0.77$$
$$\overline{1.00}.$$

This PMMA sample is predominantly, though not purely syndiotactic. According to Eqs. (10.15) and (10.16) the average sequence length is

$$l(i) = \frac{0.23}{0.18} = 1.28,$$

$$l(s) = \frac{0.77}{0.18} = 4.28.$$

Hence, the sample shows short sequences of syndiotactic PMMA which are interrupted as if by error by one or two meso placements.

Furthermore, it is possible to check whether the type of polymerisation used to produce the PMMA sample obey the simple kinetic law which presumes that each chain

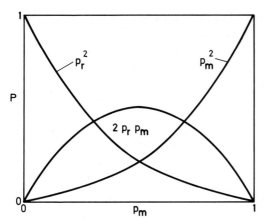

Fig. 10.6. Bernoulli statistics of polymerisation; p_m indicates the relative concentration of isotactic triads, p_r that of syndiotactic triads and $2\,P_m p_r$ that of heterotactic triads [see Eq. (10.17)]

growing step is independent on the configuration of the last segment of the growing chain. In this case, the followings (Bernoullian) statistics are valid [10].

$$P(mm)=p_m^2$$
$$P(\overleftrightarrow{mr}) =2p_m p_r$$
$$P(rr) =p_r^2 .$$

(10.17)

In Eq. (10.17) and Fig. 10.6 p_m and p_r are the probabilities for meso- and racemic placement to occur during chain growth. In the experimental example – PMMA –[16] we obtain

$$p_m=\sqrt{0.05}=0.224 ,$$
$$p_r =\sqrt{0.59}=0.768 ,$$
$$2p_m p_r = \quad 0.36 \quad \text{(experimental)},$$
$$= \quad 0.34 \quad \text{[acc. to Eq. (10.17)]} .$$

The simple kinetic law is perfectly verified in this case, as is generally true for radicalic polymerisation of PMMA [3].

Poly-α-methyl styrene (PMS)

$$\sim\!\!\text{CH}_2\!\!-\!\!\overset{\overset{\displaystyle\text{CH}_3}{|}}{\underset{\alpha}{\text{C}}}\!\!-\!\!\text{CH}_2\!\!-\!\!\sim$$

is another "simple" example that does not show any measurable ^1H spin-spin coupling in the absence of polymer-specific effects. A predominantly syndiotactic polymer (PMS st) shows peaks near $\vartheta = 0.23$ ppm $(C-CH_3)$, 1.55 ppm (CH_2) and 6.85 ppm (C_6H_5) relative to TMS [10]. Introducing meso dyads increases the heterotactic peak at 0.47 ppm and causes a signal characteristic of the (mm) triad at 0.92 ppm. Since the phenyl groups of PMS are closer to the backbone of the polymer, compared to $-O-CH_3$ in PMMA, the chemical shift of the phenyl group is affected by tacticity (6.64 ppm in PMS it).

Ordinary vinyl polymers (αH unsubstituted) do show spin coupling even in the absence of polymer specific effects:

Each αH couples with four (β) methylene protons, each methylene proton interacts with two αH. These intrinsically complicated spectra can be simplified using double resonance techniques:

– Saturation of CH_2 decouples the αH which consequently assumes the role of the $C-CH_3$ group in PMMA or PMS in probing for triads.
– Saturation of αH decouples the CH_2 group and makes the two protons inequivalent if present in an isotactic triad.

Similar simplifications may be achieved using isotopic substitution replacing either the αH or the CH_2 by deuterium. The spin resonance of ^2H is observed at a much lower magnetic field than that of ^1H, and coupling of ^2H with ^1H is weak.

An example of this type of linear polymers is PVC [2] which is obtained in "atactic" form in most polymerisation processes; the determination of the microtacticity of PVC, therefore, is of paramount importance in revealing subtle differences between samples of different origin and connections between microtacticity and properties of this polymer. The number of peaks to be expected depends on spin-spin coupling (CH_2 and αCH), configurational inequivalence of the two protons of CH_2 in isotactic PVC and the existence of three types of triads caused by αCH. Further complications may arise owing to higher order splittings if the difference in chemical shifts, $\Delta\vartheta$ equals J. In the case of vicinale protons (as in CH_2), simple 1st order spectra are expected if

$$\Delta\vartheta > 6J_{\text{vic}} . \tag{10.17}$$

The 100 MHz ^1H-NMR spectrum of radicalic PVC is shown in Fig. 10.7.

The assignment of the CH_2 and αH regions is unambiguous on account of the 2:1 ratio of the areas and the magnitude of the chemical shifts. Unfortunately, the spectrum is unsuitable for a detailed analysis because there are many unresolved peaks.

As discussed earlier, NMR spectra can be simplified by increasing the magnetic field strength and thus the chemical shift at the Hz scale; according Eq. (10.17), higher order coupling may thus be reduced to first order. Furthermore, spindecoupling and partial deuteration are suitable methods. The residual, weak $^2H/^1H$ coupling may be eliminated by deuterium decoupling. Monomeric model compounds of known configuration have to be used for comparison (Chap. 1); in the case of PVC, the dyad models are:

$$
\begin{array}{cc}
\text{CH}_3 & \text{CH}_3 \\
| & | \\
\text{H}\!-\!\overset{*}{\text{C}}\!-\!\text{Cl} & \text{H}\!-\!\overset{*}{\text{C}}\!-\!\text{Cl} \\
| & | \\
\text{CH}_2 & \text{CH}_2 \\
| & | \\
\text{H}\!-\!\overset{*}{\text{C}}\!-\!\text{Cl} & \text{Cl}\!-\!\overset{*}{\text{C}}\!-\!\text{H} \\
| & | \\
\text{CH}_3 & \text{CH}_3
\end{array}
$$

meso–2,4–dichloropentane   racemic–2,4–dichloro-
(model of m–dyad of PVC)   pentane (model of
     r–dyad of PVC)

By analogy, the steroisomers of 2,4,6-trichloroheptane are models of the isotactic (i = mm), syndiotactic (s = rr) or heterotactic (h = mr + rm = $\overset{\frown}{\text{mr}}$) triads. The NMR spectra of these dimeric and trimeric models of PVC are used as a source of chemical shifts data and coupling constants for polymer analysis, in addition to the conformational analysis of the oligomers made possible by exhaustive interpretation of the spectra. The model data can be used for a tentative synthesis of the polymer NMR spectrum.

Spin decoupling (CH$_2$) as well as deuteration of methylene protons ($\beta\beta$d$_2$ PVC) re-place the complex αH band in Fig. 10.7 ("atactic" PVC) by a "triplet" indicating the triads (mm), ($\overset{\frown}{\text{mr}}$), and (rr) (from higher to lower field), the heterotactic central peak being by far the most intense. Of course, the ^{1}H-NMR spectrum of PVC of pure tac-ticity should show only one peak (mm) or (rr). The splitting observed closely resembles that of C–CH$_3$ in PMMA, i.e. the αH serves as a "triad probe" as C–CH$_3$ does in αMe vinyl polymers. The CH$_2$ decoupled methin triple peak indicates strong con-figurational disorder; this particular PVC sample seems to be really atactic. The partly deuterated polymer ($\beta\beta$d$_2$ PVC) shows a somewhat better resolution than the spin-decoupled ordinary PVC[17]; this may also be due to its low molar mass.

The triads are assigned on the basis of model spectra (trichloroheptanes). Here, the syndiotactic (rr) model is the least shielded (ϑ higher) and the isotactic (mm) model has the most shielded αH (closer to TMS). The relative ordering is, therefore, reversed in comparison to PMMA.

The areas under the peaks yield the following relative concentrations for a radicalic (radiation-induced) PVC:

$$P(\text{rr}) = 0.328,$$

$$P(\overset{\frown}{\text{mr}}) = 0.494,$$

$$P(\text{mm}) = \frac{0.178}{1.000}.$$

The micro tacticity of this PVC sample is characterised by strong disorder, and predom-inance by very short syndiotactic sequences. According to Eqs. (10.11) to (10.14) the

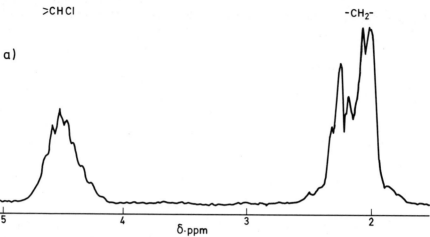

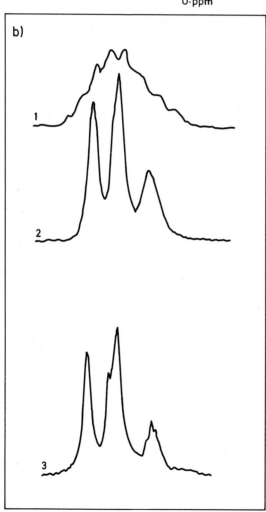

dyad distribution is given by

$$P(m) = 0.425,$$

$$P(r) = 1 - P(m) = 0.575$$

and, using (10.15) and (10.16):

$$I(i) = 0.86,$$

$$I(s) = 1,16.$$

The statistical treatment (10.17) again shows the simple Bernoulli statistics to be fulfilled.

The analyses of triads cannot go beyond the comparison with Bernoulli statistics. A more in-depth analysis has to go further, i.e. the influence of the next-nearest neighbours hat to be probed by even better spectral resolution in proceeding from triad to pentad analysis:

triad $\longrightarrow$ pentad

(2 dyads)   (4 dyads).

In order to perform such an analysis, the central proton (O) has to become mor "far-sighted", compared to that in simple triad analyses.

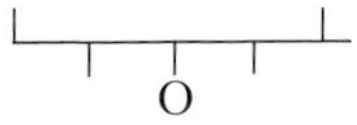

r   m   m   r

Example of a pentad (rmmr)

In the NMR spectrum, the pentad is recognised by further splitting of the central (triad) peak – in the above example the (mm) ^{1}H or C–CH$_3$ peak. Since the influence of the configuration of neighbouring units declines as a functionof the distance, the differences in chemical shift decrease, as well. There are three distinguishable pentads with a central (mm) or (rr) triad and four different pentads with a heterotactic central triad. In oligomeric $\beta\beta$-d$_2$-PVC, Cavalli et al.[17] observed splittings caused by pentad interactions and showing differences which can be attributed to the polymerisation temperature.

The next step in analysing the CH$_2$ region, probing the dyad distribution at medium resolution, involves tetrads; six distinguishable tetrads can be observed in PVC-type

---

◀ **Fig. 10.7a, b.** NMR spectra of PVC, 100 MHz, recorded in 1,2-dichlorobenzene at 120 °C (after G. Sielaff and L. Cavalli et al., redrawn from Ref. [2]. **a** total spectrum, **b** (CH region): normal spectrum (1), CH$_2$ decoupling (2), and β,β-d$_2$-PVC (3)

polymers. In this case, a specially deuterated polymer, starting from α,β-cis-dideutero vinyl chloride may provide additional information;

In this polymer, geminal ^{1}H coupling is not possible, but rather different chemical shifts are observed for (m) and (r) placements, resulting in no more then three peaks. The experimental observation of eight peaks, which can be assigned to tetrads, confirms both the assignment of triads according to the αH analysis and the validity of simple Bernoulli statistics: $P(m) = 0.44 \pm 0.02$ [compared with $P(m) = 0.425$ in triad analysis]. The small preponderance of syndiotactic configuration has been explained by electric repulsion by the Cl atoms making r-placements slightly more likely than m-placements.

By comparative studies of oligomeric model compounds, a conformational analysis can be made, indicating 90 percent trans conformation in the (r) dyads.

The NMR spectra of linear polymers burdened with bulky side groups, e.g. PVCA, may be difficult to interpret due to band broadening and shielding effects [18, 19]. The splitting of the broad CH$_2$ and CH peaks of PVCA, which significantly depends on the mode of polymerisation, has tentatively been assigned to isotactic (in cationic PVCA) and syndiotactic (in radicalic PVCA) sequences by Okamoto et al.[18]. This interpretation rests on a reasoning that differs from dyad and triad analysis. Here it is assumed that particular protons are more shielded in one conformation due to steric hindrance which again depends on the microtacticity of the chain. Model building is required in order to identify possible, sterically favourable conformations, e.g. approximately helical parts of the macromolecules.

In PVCA, the following relationships seems to be valid [18]:
- effective shielding – tightly packed conformation – isotactic helical (3/1) regions – cationic polymerisation
- less effective shielding – more open conformation – syndiotactic helical (2/1) regions – radicalic polymerisation.

This hypotheses correlates well with fluorescence spectroscopic results [20, 21]. However, NMR measurements, including ^{13}C-NMR [22] do not yet permit a final decision on the tacticity of PVCA. The poor resolution of the NMR spectra of solutions of rigid macromolecules is due to their high "internal viscosity". This problem is related to those encountered in solid polymers, as discussed in Sect. 10.4.

10.4 Broad-line NMR

In the solid state, band broadening occurs due to magnetic dipole-dipole interaction. This interaction (in the order of 10^{-3} T) decreases with the third power of the distance (10.18)

$$\Delta B \sim \frac{1}{R^3}.$$

(10.18)

The broad-line signal narrows if segmental motions become effective, comparable to dielectric and mechanical relaxations. Depending on the line broadening observed, the time scale of this experiment is of the order $\gtrsim 10^4$ Hz.

Calculations of line shapes are only possible in simple cases involving two or three interacting protons[23].

The limiting case of many interacting proton is conveniently described by a Gaussian line shape. Since Van Vleck[24], the "second moment" of the absorption curve has often been used as a measure of the line broadness, despite the fact that much information contained in the whole spectrum is lost[25].

The 2nd moment $\langle (\Delta B)^2 \rangle$ is defined according to Eq. (10.19)[23] where A is the NMR absorption observed:

$$\langle (\Delta B)^2 \rangle = \frac{\int_0^\infty A(B - B_{res})^2 \, dB}{\int_0^\infty A \, dB}$$

$$\frac{\sum_{B_1}^{B_2} A(B - B_{res})^2 \, \Delta B}{\sum_{B_1}^{B_2} A \Delta B}.$$

(10.19)

The summations are shown schematically in Fig. 10.8. The absorption (A) is weighted with respect to the distance of B from B_{res} by the integral in the numerator of Eq. (19.19). This global characterisation of band shape may in the case of multiple bands be independent of temperature, whereas the actual shape of the curve shows characteristic differences which can be interpreted in terms of molecular mobility. Figure 10.9, e.g., shows for PE that the less shielded, quasi-liquid component dramatically increases during the glass transition of the amorphous phase. A detailed analysis of this spectrum[25] shows the correspondance of the three parts of the curve to dielectric and mechanical measurements.

Molecular movement	Type of relaxation
CH_2, crystalline	α
CH_2, hindered rotation	γ
CH_2, microbrownian motion	(β, observable only in NMR).

When comparing apparent activation energies, which can be deduced from the temperature dependence of broad-line NMR as well as from relaxation measurements, the extremely broad range of relaxation times has to be taken into account. This range of relaxation times extends over roughly 10 orders of magnitude ($\log \tau \approx 0$ to -10 in seconds). The NMR line width method only covers the high frequency part, $1/\tau \gtrsim 10^5$ Hz.

Although the chemical-analytical value of broad-line NMR is small, the method is important for studying the mobility of protonated groups in solid polymers. One result of this type of NMR experiments is, e.g. that CH_3 groups in several polymers start rotating about the C–C bond at temperatures as low as -180 °C.

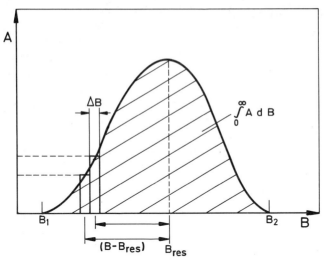

Fig. 10.8. Schematic explanation of the "2nd moment" of a NMR absorption A as a function of magnetic flux density B

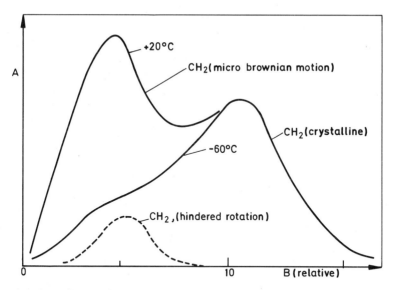

Fig. 10.9. Influence of temperature on the broad-line NMR spectra of linear PE after Bergmann[25]

In Sect. 10.8.2, a method is described which allows the measurement of ^{13}C-NMR high resolution spectra of solid polymers (Pulse-FT-NMR, combined with Magic Angle Spinning and Crosspolarisation). This method has recently been applied to ^{1}H-NMR of PET[26].

10.5 Spin-Relaxation Times

Spin resonance signals tend to saturate if no effective mechanisms exist which restore the equilibrium disturbed by absorption of electromagnetic radiation. This is a problem especially in NMR spectroscopy where the nuclear moments are weak and, hence, the possible magnetic interactions are not very pronounced as well. Spin relaxation processes depend on molecular movement, albeit in an indirect manner[27]: the relaxation (or saturation) measured is a property of the magnetic spin system which in only indirectly coupled with the molecular motions of the polymer.

Spin relaxation is described by two relaxation times which can be measured using NMR-Pulse methods, roughly analogous to the measurement of luminescence decay times:

T_1: Spin-lattice relaxation time or longitudinal relaxation time.

T_2: Spin-spin relaxation time or transversal relaxation time. (T is defined as the time needed for an excited system to reach $1/e$ of the original concentration and is called τ in all – non magnetic – spectroscopies.)

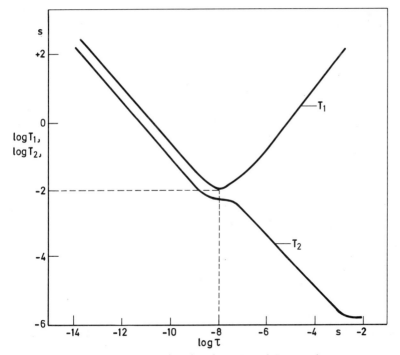

Fig. 10.10. Dependence of spin relaxation times T_1 and T_2 on polymer movements (schematic)

T_1 is due to the coupling of the nuclear spins with the "lattice" – more generally with the surrounding matrix –, coupling being caused by fluctuating magnetic fields. The fluctuations result from molecular movements; this explains qualitatively the coincidence of strong interaction and molecular movements which are of the order of the resonance frequency (about 10^8 Hz). Since the polymer movements are strongly temperature dependent, a T_1 (T) minimum indicating strong interaction can be predicted (Fig. 10.10). At very low temperature coupling with the lattice becomes very ineffective so that T_1 is very long, and in practice relaxation by traces of paramagnetic impurities, e.g. O_2, is a more effective relaxation mechanism.

T_2 is due to spin coupling with neighbouring atoms. This type of coupling is strongest at low temperature (rigid lattice) and decreases if the part of molecule which carries the absorbing group becomes mobile. In the case of very rapid movements, both T_1 and T_2 are long, i.e. coupling is ineffective.

$$T_2 \sim \frac{1}{\Delta B}.$$

(10.20)

According to Eq. (10.20), T_2 is inversely proportional to the line width (ΔB); the information contained in T_2 is therefore essentially the same as that deduced from line-width measurement [23]. The great importance of relaxation times in pulsed Fourier-transform NMR spectroscopy will be discussed in Sect. 10.8.2.

10.6 Chemically Induced Dynamic Nuclear Spin Polarisation (CIDNP)

The effect has been detected in 1967 by Bargon [28] and Ward [29]. It consists in NMR-emission or increased NMR-absorption of reaction products during radical reactions. A classical example [28] is the decomposition of benzoyl peroxide in cyclohexane at 110 °C:

– The characteristic peaks of benzoyl peroxide vanish after about 40 min
– After four minutes a C_6H_6 NMR peak appears in the emission
– This emission declines and is replaced by increased absorption after eight minutes.
  The decisive step of this reaction seems to be (10.21):

$$C_6H_5 \cdot + SH \longrightarrow C_6H_6^* + S \cdot$$

(10.21)

(SH: solvent)   (polarised)

In this case "polarised" means that a spin distribution is created which deviates from thermal equilibrium in the magnetic field of the spectrometer.

CIDNP is an important tool in studying radical reactions [30], including reactions with polymers [31].

10.7 Resumé of ¹H-NMR Spectroscopy

The greatest importance of ¹H-NMR in the polymer field (Table 10.4) lies in the determination of tacticity or rather, microtacticity of soluble polymers with pseudo-asymmetric C-atoms, the most important group belonging to vinyl polymers.

The determination of the chemical structure is a difficult task in complex polymers or mixtures; in this application, ¹³C-NMR has gained much importance due to simple spectra and the great range of chemical available. In favourable cases the preferred conformation of polymers can be calculated from experimental chemical shifts. The measurement of crystallinity and chain movements using the broad-line technique supplements other methods. The same is true for measurements of relaxation times. In the field of exciton studies, triplet excitons can in principle be studied using nuclear spin depolarisation, a method which is well known in molecular crystal studies. In complex formation, H-bonds can conveniently be studied using ¹H-NMR.

Table 10.4. Informations obtained by ^{1}H-NMR of polymers relating to structure and dynamics of polymeric systems

Structure		Dynamics	
Chemical structure	+	Movements of the chain, segments and sidegroups	+
Tacticity	+ +		
Conformation	+	Phonons	−
Crystallinity	+	Excitons	(+)
Electronic structure	+	Complex formation and related phenomena	(+)

10.8 Polymer NMR Spectroscopy of ^{13}C and Other Nuclei

10.8.1 Experimental

Nearly each element has one or more magnetic isotopes (see Table 10.1); only those isotopes which are formally build up by He nuclei have no magnetic moment: ^{4}He, ^{8}Be, ^{12}C, ^{16}O, ^{20}Ne, ^{24}Mg, ^{28}Si etc. In principle, therefore, it should be possible to use nearly all elements in NMR spectroscopy, although for reasons of sensitivity only ^{19}F can be measured on conventional continuous mode spectrometers (see Table 10.1). The main causes for low NMR sensitivity of an isotope are low natural abundance and small magnetic moment. Furthermore, the nuclear quadrupole moment of certain nuclei (e.g. ^{14}N, ^{17}O, ^{33}S) broadens the spectra so that the effective sensitivity of measuring these isotopes is decreased. The disadvantage of low sensitivity may well be overcompensated by very weak spin-spin coupling due to low abundance (only very few pairs of the same isotope such as ^{13}C-^{13}C) once a method for recording spectra with a high degree of sensitivity is found. Such a method has become available (PFT-NMR) which combines excitation by short (μs) rf pulses with Fourier-transformation of the signals obtained. On the other hand, the sensitivity of conventional continuous mode

Table 10.5. Comparison between CW-NMR and PFT-NMR Spectroscopy

	CW-NMR	PFT-NMR
rf-Radiation	Continuous	Pulsed
Intensity (power)	Small	High
Spectral purity of radiation	Monochromatic	Polychromatic
Magnetic field	Variable (in field sweep operation)	Constant
Signal	$I(v)$ or $I(B)$	$I(t)$ FT $\downarrow$ $I(v)$ or $I(B)$
Signal averaging	Time consuming	Rapid
Computer	Peripheral	Integral part
Nuclei	^{1}H, ^{19}F	All nuclei if natural abundance is sufficient
Solid state	Broad-line NMR	High resolution possible (MAS-CP)

recording cannot be substantially increased due to saturation if the excitation power is increased (in addition to limitations in time and field stability during SN improvement by averaging).

Using the pulse method, one spectrum per pulse is stored, the minimal repetition time being five spin-lattice relaxation time ($5\ T_1$). If non equivalent nuclei are present, the longest T_1 has to be used to calculate the repitition rate. If T_1 is small, a large number of spectra can be stored and used for computational SN averaging.

The rf excitation pulses in PFT-NMR cover a frequency range which is at least as broad as the spectrum expected. They, therefore, contain all frequencies needed fo a full NMR spectrum to be recorded if the magnetic field is kept constant during the experiment. Owing to the shortness of the pulses, the power level of the exciting radiation can be much higher than in conventional CW-NMR spectroscopy. The magnetic nuclei of the sample to which the apparatus has been tuned, e.g. ^{13}C, ^{15}N, etc., absorb their characteristic frequencies from the radiation offered to them whereby the magnetic moments are aligned with respect to the external magnetic field. The signal $I\,(t)$ resulting after the pulse from reorientation of the excited spins corresponds to the interferogram in FT-IR spectroscopy, i.e. it contains the full spectral information in a coded form. Using the Fourier transformation technique, the signal $I\,(t)$ is computationally transformed into the NMR spectrum $I\,(v)$ or $I\,(B)$. The signal $I\,(t)$ is called FID (free induced decay).

Modern PFT-NMR spectrometers allow nearly all magnetic nuclei to be measured, each nucleus being characterised by a resonance (at a given field), which is slightly modified by the chemical shift. The frequency range of rf excitation has to be chosen for each magnetic nucleus to be measured in such a way that the resonance frequency plus the chemical shift range is contained in the spectral width. Strong magnets of the superconducting coil type furnish a magnetic flux density of $B \approx 7$ Tesla or even higher; the resonance frequency of most nuclei at 7 T lies between $v_0 \approx 10$ to 100 MHz (compared to about 300 MHz for ^1H and ^{19}F). It is necessary that the pulses can be chosen in defined sequences or programmes and that other nuclei, especially ^1H can be decoupled. Table 10.5 shows a comparison of the most important differences between CW- and PFT-NMR spectroscopy. The PFT-spectrometer can be used not only for recording NMR spectra but also for performing relaxation measurements, e.g. by saturation experiments using rapid pulse sequences. In this respect, the PFT technique is also advantageous for measuring NMR spectra of ^1H and ^{19}F. Furthermore, dilute solutions can be measured (important for polymers) and kinetic measurements can be performed in the time region $\gtrsim 10^{-6}$ s.

10.8.2 Solid Polymer ^{13}C-NMR Spectroscopy

The most important nucleus in polymer spectroscopy is ^{13}C since [32, 33]:
- Nearly all polymers show C-atoms in the main chain, or at least in side-groups.
- The range of chemical shifts is very large (200 ppm).
- Spin-spin coupling between neighbouring atoms (^{13}C-^{13}C) – not to be eliminated by spin decoupling – is very weak owing to the low natural abundance (Table 10.1); the probability of such contacts is about 10^{-4}.

The informations obtained by ^{13}C-NMR is equally important in organic chemistry and in polymer research.

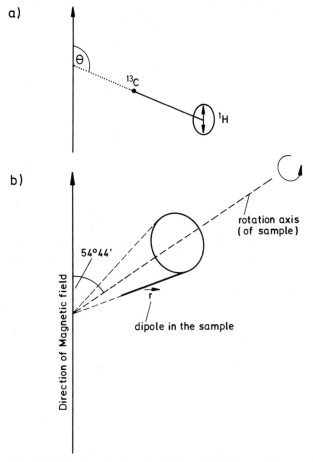

Fig. 10.11. Magic angle spinning (MAS), **a** Definition of the angle θ, **b** Schematic view of the MAS experiment; **r** represents a dipole of a random sample. All possible **r** vectors by rapid rotation about the axis indicated on the average include the "magic angle" with the magnetic field

The technique leading to well resolved solid state spectra (MAS) has first been explored by polymer spectroscopists[34-36]. In general, solid state NMR spectra are broad, as discussed in Sect. 10.4 for ^{1}H-NMR. The main reasons for low intensity, broad ^{13}C-NMR spectra in solids are[37]:
a) Anisotropic magnetic dipole-dipole interaction, mostly ^{13}C-^{1}H,
b) Anisotropy of chemical shift,
c) Long relaxation time T_1 (several minutes for ^{13}C in solids) – limiting the repetition frequency.

Restriction (c) is "merely" a problem of sensitivity, whereas (a) and (b) are fundamental restrictions. The experimental solution of high resolution in solid state ^{13}C-NMR requires the following combination of techniques[48]:
1) High power ^{1}H-decoupling,
2) Rotation of the sample about the "magic angle" = magic angle spinning (MAS),
3) Cross polarisation (CP).

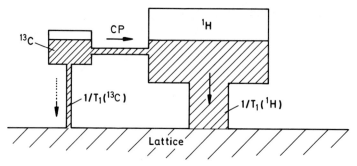

Fig. 10.12. Effect of Cross Polarisation (hydrostatic analogue)

High power ^{1}H-decoupling (1) solves the problem of dipolar broadening (a). This could, in principle, be solved by (2) at very high rotation speed. MAS (2) is based on the fact that both intrinsic broadening mechanisms (a) and (b) contain the term $(3 \cos^2 \theta - 1)$. This is visualised in Fig. 10.11 for ^{13}C-^{1}H interaction. The above term is zero at $\theta = 54°44'$ relative to the external magnetic field. If the sample is rapidly rotated at this angle, as indicated in Fig. 10.11, the average θ for all vectors (dipoles) in the sample is the magic angle and all dipole interactions as well as chemical shift anisotropics cancel. The rotation has to be faster than the frequency of the broadening mechanism to be eliminated; the resonance condition for ^{13}C is (10.22)

$$J_0(\text{MHz}) \approx B_0(\text{kG}) . \tag{10.22}$$

A 10 ppm broadening (at $B_0 \approx 7$ $T = 70$ kG) would correspond to 700 Hz; a spinning frequency of several kHz, therefore, eliminates this broadening.

Cross polarisation (3) overcomes the long T_1 of ^{13}C by a sequence of suitable pulses which transfer the polarisation from slowly decaying ^{13}C to the much faster relaxing ^{1}H system. This principle of CP is shown in Fig. 10.12 by means of a hydrostatic analogue. The transfer of polarisation from the ^{13}C to ^{1}H system is induced by satisfying the condition (10.23)

$$\omega(^{13}\text{C}) = \omega(^1\text{H}) \tag{10.23}$$

essentially by simultaneous irradiation with both ^{13}C and ^{1}H resonance frequencies.

Using the CP-MAS method, even insoluble, e.g. cross linked polymers can be studied with a resolution high enough for chemical analysis. The only prerequisite is the presence of a magnetic nucleus in the sample, usually ^{13}C, in small concentration.

CP-MAS is especially important for many commercial plastics which often are insoluble. As an example [36], epoxides have been investigated, consisting of bis-phenol A-diglycidylether as a bifunctional monomer (DGEBA):

DGEBA

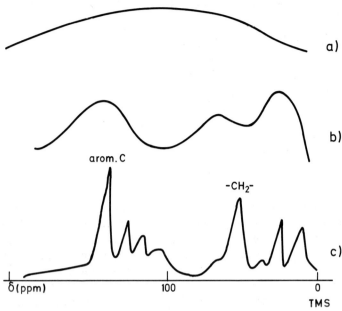

Fig. 10.13a–c. Solid state NMR of DGEBA converted by 5% piperidine into an insoluble resin; **a** ^{13}C-broad line spectrum, **b** 60 kHz ^{1}H decoupling, **c** 2.2 kHz MAS

This monomer is converted by amines into insoluble resins. The increasing resolution from the ^{13}C broad line to the well resolved MAS spectrum can be seen in Fig. 10.13.

The application of NMR to solid polymers has recently been reviewed by McBrierty and Douglas[49]. The use of ^{13}C-MSA for conformational studies of solid polymers has been demonstrated by Cantow and coworkers[50, 51].

10.8.3 High Resolution ^{13}C-NMR Spectroscopy of Polymer Solution

The usefulness of ^{13}C-NMR and the advantages compared to ^{1}H-NMR are due to
large range of chemical shifts enabling the resolution of chemically very similar structural units nearly complete absence of ^{13}C-^{13}C coupling broad band decoupling of ^{13}C-^{1}H coupling avoidance of saturation using the pulse technique.

Polymer solutions, showing a high degree of segment mobility (in order to prevent dipolar broadening) are highly suitable for recording ^{13}C-NMR spectra. The main information which can be obtained from these spectra includes
identification of specific polymers[38], identification of groups in a polymer, branching of macromolecules, analysis of sequences[33, 52] in copolymers, distinction between block- and random-copolymers, analysis of (micro-)tacticity[33].

Polymers and groups in the polymer chain are identified by comparison with standard spectra, now available for ^{13}C-NMR[38], tables of chemical shift data[39], and comparison with spectra of monomeric model compounds recorded under exactly the same conditions with regard to solvent, temperature, B_0, etc. as the polymer samples.

Branching of carbon chains is a most important subject of ^{13}C-NMR studies[40]. The diamagnetic shielding by nearest (α), next to nearest (β) and even γ C-atoms of the bran-

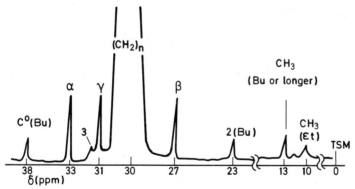

Fig. 10.14. Typical ^{13}C-NMR spectrum of LDPE in solution (simplified)

ching point C (o) results in a broad range of chemical shift.

$$- {}^{13}\overset{\alpha}{C}{}^o - \overset{\beta}{C} - \overset{\gamma}{C} - C -$$

C-atoms in α and β position cause strong deshielding (increase of ϑ) of about 9 ppm/ C-atom. C-atoms in γ position, on the other hand, have a shielding effect and thus decrease the chemical shift observed (exception: quarternary γ C-atoms).

These empirical rules which have been obtained from a great number of monomeric models allow us to construct theoretical spectra for different types of branching, including the endgroups which can be observed in low molar mass polymers.

LDPE is one of the commercially most important branched polymers, produced by radicalic initiation of ethene at high pressure. The ideal structure of PE, a chain consisting of CH_2 groups, which is approximately realised in HDPE, gives a ^{13}C-NMR signal after ^{1}H-decoupling at $\vartheta \approx 30$ ppm relative to TMS.

This peak is also the strongest one in solution spectra of LDPE and indicates the unperturbed regions of the main chain and of long-chain branches which might be present.

$$\text{n-butyl branch} \begin{cases} CH_3 & \longleftarrow \quad 1 \\ | \\ CH_2 & \longleftarrow \quad 2 \\ | \\ CH_2 & \longleftarrow \quad 3 \\ | \\ CH_2 & \longleftarrow \quad 4 \end{cases}$$

$$\underset{(\alpha)}{CH_3} - \underset{(\beta)}{CH_2} - \underset{(\gamma)}{CH_2} - CH_2 - \overset{30 \text{ ppm}}{(CH_2)_n} - \underset{\gamma}{CH_2} - \underset{\beta}{CH_2} - \underset{\alpha}{CH_2} - \overset{|}{C^o} - \underset{\alpha}{CH_2} - \underset{\beta}{CH_2} - \underset{\gamma}{CH_2}$$

Main chain (LDPE)

The best shielded C-atom as a rule is located in CH_3-end groups (Fig. 10.14). The branching point (C^0) "feels" the slightest shielding due to the combined deshielding effects of main chain plus branch $\alpha + \beta$ C-atoms. The spectrum shown in Fig. 10.14 essentially corresponds to the one calculated for n-butyl branches[41, 42] and confirms the "back-biting" hypothesis of branching (10.24), postulated before ^{13}C-NMR spectroscopy was introduced:

$$-CH_2-\underset{\underset{CH_2}{|}}{\overset{\overset{H}{|}}{C}}\cdots \quad \bullet CH_2 \quad \longrightarrow \quad -CH_2-\underset{\underset{CH_2}{|}}{\overset{\overset{H}{|}}{C}}\bullet$$

$$\xrightarrow{+\,C_2H_4} \quad -CH_2-\underset{\underset{CH_2}{\underset{|}{|}}}{\overset{\overset{H}{|}}{C}}-CH_2-CH_2^\bullet \qquad \text{etc.} \qquad (10.24)$$

Formation of n–butyl branches in LDPE by "back–biting"
of the growing chain end

Polyethene and other polymers without pseudo-asymmetric C-atoms are especially suited for NMR branching studies since splittings due to differences in tacticity cannot occur. The spectrum of native PVC is too complicated for studying the branching pattern; a total reduction of PVC using $LiAlH_4$ yields, without chain splitting, PE of the same branching as the original PVC[40]. In this case the ^{13}C-NMR spectrum shows mostly CH_3 bands ($\vartheta \approx 20$ ppm relative to TMS). A quantiative evaluation yields about 3 CH_3 groups per 1,000 C-atoms. This short chain branching may occur at head-to-head placements by shift of a Cl-atom (20.25):

$$-CH_2-\underset{\underset{Cl}{|}}{CH}-\underset{\underset{Cl}{|}}{CH}-CH_2^\bullet \quad \longrightarrow \quad -CH_2-\underset{\underset{Cl}{|}}{CH}-\underset{\underset{CH_2Cl}{|}}{\overset{\overset{H}{|}}{C}}\bullet$$

head–to–head

$$\xrightarrow{+\,C_2H_3Cl} \quad -CH_2-\underset{\underset{Cl}{|}}{CH}-\underset{\underset{CH_2Cl}{|}}{CH}-CH_2-\underset{\underset{Cl}{|}}{\overset{\overset{H}{|}}{C}}\bullet \qquad \text{etc.} \qquad (10.25)$$

The CH_2Cl groups, after reduction, are identified as CH_3 groups[40].

Another important application of ^{13}C-NMR is the study of copolymers[33, 52]:

– $(A)_n$–$(B)_m$ – block-copolymers

- ABABAB – alternating copolymers
- AABABBA– statistical copolymer (irregular)

Basically, the analytical problems are similar to those encountered when analysing the microtacticity of polymers[33], see Sect. 10.3.1. In order to perform a quantitative determination, the chemical shifts of ^{13}C-atoms within sequences of one comonomer and at the links have to be different (10.26).

$$\vartheta^{13}C(A) \neq \vartheta^{13}C(B) \neq \vartheta^{13}C(AB) . \tag{10.26}$$

Since the influence of neighbouring atoms extends over three atoms "to the right" and "to the left" of $^{13}C^0$, each C-atom in vinyl polymers can "feel" both adjoining monomers (A or B).

The statistical evaluation according to dyads, triads etc. is similar to tacticity analysis. This evaluation is treated in extenso in Randall's monograph[33].

As an examle, polybutadiene, formally a homopolymer, is in fact a copolymer due to 1,3 and 1,2 addition:

$$CH_2=CH-CH=CH_2 \qquad\qquad \text{1,3-butadiene}$$
$$-CH_2-CH=CH-CH_2-CH_2-CH=CH_2-CH_2- \quad \text{1,4 addition}$$
$$\begin{array}{c} -CH_2-CH-CH_2-CH- \\ | \qquad\quad | \\ CH=CH_2 \quad CH=CH_2 \end{array} \qquad \text{1,2 addition}$$

Hydrogenation of polybutadiene transforms this polymer to PE with ethyl branches. The ^{13}C-NMR spectrum shows 19 peaks, $\vartheta = 10$ to 40 ppm. The statistical analyses yields the relative number of branches and the distribution of segment lengths.

In partially hydrogenated samples, the degree of reduction can be measured. This can be taken as a typical example of an application of ^{13}C-NMR in industrial practice [43]. Here, as in the case of sequence analysis, the correct assignment of all peaks is decisive. A rough assignment is possible using tables; small splittings have to be explained using model spectra, different polymers, etc. However, absolute certainty can not be achieved due to conformational effects on the chemical shifts. Deviations between model and polymer and between different polymers are possible. Therefore, different peaks of a spectrum have to be evaluated and the consistency of the results has to be checked.

10.8.4 Other Nuclei

Because of the versatility of modern PFT-NMR spectrometers, other nuclei present in polymers will be increasingly studied. As an example, Kricheldorf[44] applied ^{15}N-NMR to polymer problems. According to this work, homo- and copoly urethanes can be studied using ^{15}N-NMR, since N is part of the main chain and structural differences in neighbouring basis units shift the resonance signal of ^{15}N.

In linear PU of the type[44]

$$\left[-NH-(CH_2)_n-NH-\underset{\underset{O}{\|}}{C}- \right]_n$$

the shielding of ^{15}N decreases with increasing n, i.e. $\vartheta\ ^{15}$N increases.

Table 10.6. Informations obtained by ^{13}C-NMR of Polymers relating to structure and dynamics of polymeric systems

Structure		Dynamics	
Chemical structure	$++$	Movements of the chain, segments and sidegroups	$+$
Tacticity	$+$		
Conformation	$+$	Phonons	$-$
Crystallinity	$(-)$	Excitons	$-$
Electronic structure	$-$	Complex formation and related phenomena	$-$

^{2}H-NMR is used extensively for studying relaxation processes in deuterated polymers by Sillescu and coworkers [45, 53, 54]. The advantage of using ^{2}H- instead of ^{1}H-NMR is a different relaxation (coupling) mechanism [45] which facilitates the interpretation of the experiments.

10.9 Résumé

The focus of ^{13}C-NMR in the polymer field (Table 10.6) is on chemical analysis, where the greater simplicity of the spectra and the larger chemical shift range, compared to ^{1}H-NMR, makes this technique unique. Copolymers and polymers of different tacticity can also be analysed. For the first time, high resolution NMR spectra of solids have been obtained using the CP-MAS technique allowing structural identification as well as conformational analysis of macromolecules in the solid state. This is especially true for PFT-NMR; spin relaxation, however, is only indirectly coupled with molecular motion. This must be kept in mind when evaluating NMR relaxation experiments of polymers.

References

1. Ingram, J.E.: Spectroscopy at Radio and Microwave Frequencies, 2nd Ed., London: Butterworth 1967
2. Klesper, E., Sielaff, G. in Hummel, D.O. (ed.): Polymer Spectroscopy, Weinheim: Verlag Chemie 1974, p. 189
3. Bovey, F.A.: High Resolution NMR of Macromolecules, New York: Academic Press 1972
4. Levins, J.N.: Molecular Spectroscopy, New York: Wiley 1975, p. 331
5. Tables of chemical shifts are reproduced in recent editions of the Handbook of Chemistry and Physics, CRC Press
6. Phillips, W.D.: J. Chem. Phys. *23*, 1363 (1955)
7. Bargon, J., Fischer, H., Johnsen, U.: Z. Naturforsch. *22a*, 1551 (1967)
8. Richard, C., Granger, P.: *NMR* Vol. 8, Berlin, Heidelberg, New York: Springer 1974
9. Sadtler Commercial Spectra, Proton NMR-Polymers 60 MHz (2 Vol.); Polymers 100 MHz (1 Vol.); 300 spectra per vol., London: Heyden 1980
10. Johnsen, U.: Kolloid Z.v.Z. Polymer *210*, 1 (1966)
11. Woodbrey, J.C., in: Ketley, A.D. (ed.): The Stereo-chemistry of Macromolecules, Vol. 3, New York: Dekker 1968
12. Bovey, F.A., Tiers, G.V.D.: J. Pol. Sci. *44*, 173 (1960)
13. Johnsen, U., Tessmar, K.: Kolloid Z.v.Z. Polymere *168*, 160 (1960)
14. Nishioka, A., Watanabe, H., Yamaguchi, I., Shimizu, H.: J. Pol. Sci. *45*, 232 (1960)

D. Spin-Resonance Spectroscopy

15. Natta, G.: J. Pol. Sci. *16*, 143 (1955)
16. Spectrum by Dr. Wendisch, in: Hoffmann, M., Krämer, H., Kuhn, R.: Polymeranalytik I, Thieme Taschenlehrbuch B 4 1977
17. Cavalli, L., Borsini, G.S., Carraro, G., Confalonieri, G.: J. Pol. Sci. A1, 8, 801 (1970)
18. Okamoto, K.-I., Yamada, M., Itaya, A., Kimura, T., Kusabayashi, S.: Macromolecules *9*, 645 (1976)
19. Williams, D.J.: Macromolecules *3*, 602 (1970)
20. Rippen, G., Kaufmann, G., Klöpffer, W.: Chem. Phys. *52*, 165 (1980)
21. Itaya, A., Okamoto, K.-I., Kusabayashi, S.: Bull. Chem. Soc. Japan *49*, 2037 (1976)
22. Kawamura, T., Matsuzaki, K.: Makromol. Chem. *179*, 1003 (1978)
23. Kosfeld, R., v. Mylius, U., in: Diehl, P., Fluck, E., Kosfeld, R. (eds.): NMR, Vol. 4, Berlin, Heidelberg, New York: Springer 1971, p. 181
24. van Vleck, J.H.: Phys. Rev. *74*, 1168 (1948)
25. Bergmann, K.: see Ref. [23], p. 233
26. Cheuny, T.T.P., Gerstein, B.C., Ryan, C.M., Taylor, R.E.: J. Chem. Phys. *73*, 6059 (1980)
27. Broecker, H.Ch.: Third Int. Seminar on Polymer Physics, High Tatra, April 25 to May 1, 1982
28. Bargon, J., Fischer, H., Johnson, U.: Z. Naturforsch. *22a*, 1551 (1967)
29. Ward, H.R.: J. Am Chem. Soc. *89*, 5517 (1967)
30. Richard, C., Granger, P.: NMR, Vol. 8. Chemically Induced Dynamic Nuclear and Electron Polarizations – CIDNP and CIDEP, Berlin, Heidelberg, New York: Springer 1974
31. Bargon, J.: CIDNP During the Photolysis of Poly (Methyl Isopropenyl Ketone) and its Model Compounds in Solution. Int. Symposium on Degradation and Stabilization of Polymers, Brussels 11–1, September 1984
32. Pasika, W.M. (ed.): Carbon-13 NMR in Polymer Science, ACS Symposium Series, Vol. 103, Washington 1979
33. Randall, J.C.: Polymer Sequence Determination Carbon-13 Method, New York: Academic Press 1977
34. Schaefer, J., Stejskal, E.O.: J. Am. Chem. Soc. *98*, 1031 (1976)
35. Schaefer, J., Stejskal, E.O., Buchdall, R.: Macromolecules *10*, 384 (1977)
36. Garroway, A.N., Moritz, W.B., Resing, H.A., in Ref. [32], p. 67
37. Lyerla, J.R., in: Fava, R.A. (ed.): Method of Experimental Physics, Vol. 16 Polymers, Part A, New York: Academic Press 1980, p. 241
38. Carbon 13 NMR of Monomers & Polymers, Sadtler Research Lab. (1979)
39. See Ref. [5]
40. Bovey, F.A.: High-Resolution Carbon 13 Studies of Polymer Structure, in: Ivin, K.J. (ed.): Structural Studies of Macromolecules by Spectroscopic Methods, London: Wiley 1976, p. 181
41. Dorman, D.E., Otaka, E.P., Bovey, F.A.: Macromolecules *25*, 574 (1972)
42. Randall, J.C.: J. Pol. Sci., Pol. Phys. Ed. *11*, 275 (1973)
43. Arendt, G.: Personal communication
44. Kricheldorf, H.R.: J. Macromol. Sci. Chem. A*14*, 959 (1980)
45. Collignon, J.L., Sillescu, H., Spiess, H.W.: Colloid Pol. Sci. *259*, 220 (1981)
46. Harris, R.K., Packer, K.J., Say, B.J.: Makromol. Chem. Suppl. *4*, 117 (1981)
47. Becker, E.D.: High Resolution NMR-Theory and Chemical Applications, 2nd. Ed., New York: Academic Press 1980
48. Mehring, M.: Principles of High Resolution NMR in Solids, Berlin, Heidelberg, New York, Tokyo: Springer 1983
49. McBrierty, V.J., Douglas, D.C.: J. Pol. Sci. Macromol. Rev. *16*, 295 (1981)
50. Möller, M., Cantow, H.-J.: Pol. Bull. *5*, 119 (1981)
51. Gronski, W., Hasenschindl, A., Limbach, H.H., Möller, M., Cantor, H.-J.: Pol. Bull. *6*, 93 (1981)
52. Hsieh, E.T., Randall, J.C.: Macromolecules *15*, 353 (1982)
53. Lindner, P., Rössler, E., Sillescu, H.: Macromol. Chem. *182*, 3653 (1981)
54. Sillescu, H.: Pure Appl. Chem. *54*, 619 (1982)

Part E. Conclusion and Appendices

Conclusion

Several of the most important methods of polymer characterisation and identification are spectroscopic methods; among them, mIR and NMR are useful for all polymers, other methods, such as nUV, for certain types of polymers only. Spectroscopy complements the "classical" methods of polymer research, such as X-ray and light scattering, osmometry, rheology, dielectric and mechanical relaxation, X-ray diffraction, etc.

The most fascinating aspect of polymer spectroscopy, as defined in Chap. 1, is that of a "dialogue" at the molecular level, where molecular in most cases means that one or a few monomer units are involved, rather than the macromolecule as a whole. Neighbouring groups – in extreme cases the whole polymer molecule – may influence the signals obtained (polymer aspect). However, the distances involved in these interaction rarely surpass 1 to 10 nm, a fact which has been called the "short-sightedness" of most spectroscopic methods at several occasions in this book.

The dialogue, if we stick to this metaphor for a moment longer, is, of course, a coded one. Decoding in most cases is performed empirically by comparison with the spectra of monomers, oligomers, related polymers etc. There are other spectra, however, which can only be understood on the basis of calculations.

The most extensive application of polymer spectroscopy, especially in practical and routine work, seems to be the chemical analysis of polymers including subtle structural differences such as the microtacticity. In addition, spectroscopy provided plenty of chemical and physical information about polymers which so far has been derived only for relatively simple and commercially important polymers. It was perhaps the most important aim of this book to point out these exciting yet not fully exploited possibilities of polymer spectroscopy.

Recent experimental developments are connected with Fouriertransform techniques in the long wavelength region (PFT-NMR and FTIR), whereas in the short wavelength region ESCA has been the main new development in the last decade. In the fUV, together with nIR, the most neglected field of polymer spectroscopy, a change seems possible owing to Synchroton radiation which has become available in some laboratories connected to electron accelerators. In nUV/VIS spectroscopy it is especially the investigation of dissolved macromolecules which has recently found much attention in polymer research. In this area, nano- to picosecond spectroscopy has been opening the door to studies of extremely fast kinetics.

E. Conclusion and Appendices

Finally, it should be pointed out that a close cooperation between polymer spectroscopists and their colleagues working on the preparation of polymers is needed and in many cases has to be improved. The purity and homogeneity of the samples investigated is an important criterion in many experiments in this field, especially at the level of basic research. However, even in polymer identification or functional group analysis, high-purity and well defined reference samples are desirable.

Appendix 1

Table of Polymers

Chemical name	Abbreviation	Section
Deoxyribonucleic acid	DNA	3.2.5
High density-poly (ethene) (predominantly linear)	HDPE	2.5.4; 3.1; 6.4; 6.5; 7.7.3; 10.4
Low density-poly (ethene) (branched)	LDPE	2.5.4; 3.1; 6.4; 7.5.4; 10.8.3
Poly (acrylic acid)		9.4
Poly (acrylonitrile)	PAN	7.4.2.4; 7.58
Poly (acrylonitrile-co-methyl-methacrylate)	PAN/MMA	7.5.2
Polyamide (s)	PA	3.2.4; 7.4.2.5; 9.8
Polyamide (66)	PA-66	4.4.5
Poly-(butadiene)		7.4.3; 7.5.5; 10.8.3
Poly (1-butene)	PB	2.4.1; 2.5.4
Poly (caprolactam)	PA-6	7.4.2.5; 9.5
Poly (p-chlorostyrene)	pClPS	9.8
Poly epoxides based on the di-glycidylether of bisphenol A	DGEBA	10.8.2
Poly (ethene)	PE	2.4.1; 3.1; 5.5; 6.4; 6.5; 7.5.6; 9.6; 10.8.3
Poly (ethene oxide)	PEO (POE)	3.1; 9.8
Poly (ethene terephthalate)	PET	2.4.1; 3.2.4; 7.5.6; 7.7.2
Poly (p-fluorostyrene)	pFPS	9.8
Poly (isobutene)	PIB	7.4.2.2; 9.6
Poly (isoprene)		7.5.5
Poly (α-methylstyrene)	PMS	10.3.2
Poly (methyl methacrylate)	PMMA	3; 7.4.2.1; 9.8; 10.3.2
Poly (methyl methacrylate-co-styrene)	PMMA/S	4.4.2
Poly (oxymethylene)	POM	2.5.4; 7.4.3; 7.7.4
Poly (propene)	PP	2.4.1; 2.5.4; 6.4; 7.4.2.2; 7.4.3; 7.5
Poly (styrene)	PS	2.4.1; 3.1; 3.2; 4.4.4; 4.5; 9.8
Poly (styrene-co-vinyl carbazole)	PS/VCA	4.4.2; 4.5.2
Poly (styrene-co-vinyl naphthalene)	PS/VN	4.4.2
Poly (styrene sulphonic acid)		7.5.8
Poly (tetrafluoro ethene-co-ethene)	PTFE/E	2.4.3; 2.4.4
Poly (tetrafluoro ethene)	PTFE	2.4.1; 2.4.2; 2.4.4; 6.4; 7.7.3; 7.7.4
Polyurethane(s)	PU	3.2.4; 10.8.4
Poly(vinylacetate)	PVAC	3.1; 7.4.2.3
Poly(vinylalcohol)	PVOH	2.4.1; 7.4.2.3
Poly(N-vinyl carbazole)	PVCA	3.2.4; 4.4.4; 4.5.1; 9.9; 10.3.2
Poly(vinylchloride)	PVC	2.4.1; 2.5.4; 10.3.2; 10.8.3
Poly(vinylfluoride)	PVF	2.4.1
Poly(vinylidene fluorid)	PVF$_2$	7.7.2
Poly(vinylidene chloride)	PVCl$_2$	5.5; 7.4.2
Poly (1-vinyl naphthalene)	P1VN	9.1
Poly (2-vinyl naphthalene)	P2VN	4.4.5
Polymer of bis(p-toluene-sulpho-nate) of 2,4-hexadiine-1,6-diol	TSHD (PTS)	6.6; 9.7

Appendix 2

List of Abbreviations

Roman and Italics Letters

		Unit[1]
a	Lattice constant	nm
a	Electron-nuclear hyperfine coupling constant (ESR)	G, **T, Hz**
a.f.	Alternating frequency	
AFC	*A*utomatic *F*requency *C*ontrol	
A_g	Electron affinity of free molecule or basic unit of polymer	eV
A_g^*	Electron affinity of free molecule in an excited electronic state	eV
A_s	Electron affinity of solids	eV
at	Atactic	
ATR	*A*ttenuated *T*otal *R*eflection	
B	Magnetic flux density or induction	**T (Vsm^{-2})**, G
b	Bending vibration	
$B(v')$	Spectral intensity distribution of IR radiation (FTIR)	
br	Breathing vibration	
c	Velocity of light	**ms^{-1}**
c	Molar concentration	mol basic unit l^{-1} (1 l = 1 dm^3)
c'	Concentration (mass/volume)	g l^{-1} (**kgm^{-3}**)
CARS	*C*oherent *A*ntistokes *R*aman *S*pectroscopy	
CB	Conduction band	
C_G, C_E	Concentration (mol ratio) of guest molecules or excimer-forming sites	mol · (mol basic unit)$^{-1}$
CIDNP	*C*hemically *I*nduced *D*ynamic *N*uclear *S*pin Polarisation	
C_n	n fold axis of symmetry	
CT	*C*harge *T*ransfer (inter- or intramolecular CT band, CT (donor acceptor) complex)	
CW	*C*ontinuous *W*ave	
D, D^*	First order zero field splitting parameter (triplets) and mean zero field splitting parameter	cm^{-1} GHz
1D, 2D, 3D	One, two, three dimensional (lattice)	

---

[1] SI basic unit in bold letters

d	Thickness of polymer film or solution in optical absorption measurements	
d	Deformation vibration	
E_F	Fermi energy	eV
E_{kin}	Kinetic energy	**J**, eV
E_{opt}	Spectroscopically measured absorption edge	eV, cm^{-1}
E_{photo}	Threshold of intrinsic photo-current in photo-action spectra	eV, cm^{-1}
E_{pot}	Potential energy	**J**, eV
ESCA	Electron Spectroscopy for Chemical Analysis (or Applications)	
ESR	Electron Spin Resonance	
$f_D(v')$	Fluorescence function (rel. spectral distribution) of the energy donor	
FID	Free Induced Decay (in FT-NMR)	
fIR	far infrared	
$f_{k,n}$	Oscillator strength of a transition (between states k and n, $k < n$)	
FMIR	Frustrated Multiple Internal Reflection	
fUV	far UV	
G	Gauß, 10^{-4} Tesla	
g	g-factor (g_e of electrons, g_N of nuclei, g_P of proton)	
GPC	Gel Permeation Chromatography	
H	Magnetic field strength	**A m^{-1}**
h	Planck's constant	**Js**
$\hbar$	$h/2\pi$	**Js**
hv	Photon energy; in chemical equations: photochemical reaction	eV photon^{-1} **J Einstein^{-1}** [2]
I, I_0	Intensity of radiation, intensity of reference beam	**W m^{-2}** photons cm^{-2} Einstein cm^{-2}
i	Centre of symmetry	
i	iso	
i	(as subscript) non-radiative	
ic	Internal conversion	
$I_1, I_2 ... I_n$	Ionisation energy of the highest, second highest... molecular orbital	eV
I_g	$(= I_1)$ Ionisation energy of free molecule or basic unit of polymer	eV

---

2 1 Einstein = 1 mol photons

I_g^*	Ionisation energy of free molecule in excited electronic state (S_1 or T_1)	eV
INS	*I*nelastic *N*eutron *S*cattering	
isc	*i*nter-*s*ystem *c*rossing	
I_s	Intensity of scattered light (radiation)	**W m^{-2}**
it	Isotactic	
$\dfrac{I_E}{I_M}$	Intensity ratio of excimer to monomer fluorescence	
J, J'	Spin-spin coupling constant	**s^{-1} (Hz)**
J_A	Exchange interaction energy	eV
k	Boltzmann's constant	**J K^{-1} molecule^{-1}**
k	Force constant	**kg s^{-2}**
k	"Wave number" of electrons and holes in the reciprocal lattice, lattice constant a	$\dfrac{1}{a}$
k_e, k_e'	Rate constants of radiative deactivation (of fluorescence or phosphorescence)	**s^{-1}**
k_{ic}	Rate constant of internal conversion (mostly $S_1 \to S_0$)	**s^{-1}**
k_{isc} k_{isc}'	Rate constants of intersystem crossing ($S \to T$ or $T \to S$)	**s^{-1}**
kT	Average thermal energy at temperature T (K)	eV molecule^{-1}
L	Length (of vibrating polymer section)	**m**, nm
M	Molar mass (formerly: molecular weight)	g mol^{-1}
$\bar{M}(\bar{M}_n; \bar{M}_w)$	Average molar mass; with index n: number average; with index w: weight average	
m	Meso (configuration of dyad)	
m	Medium (IR absorption intensity)	
MAS	*M*agic *A*ngle *S*pinning (in solid state NMR)	
M, D, P	Monomer, Dimer, Polymer and corresponding radicals and radical-ions	
m_e	Mass of the electron	**kg**
mIR	*m*edium *i*nfra*r*ed	
M$_{k,n}$	Transition dipole moment (between states k and n)	D (Debey, 10^{-18} esu) **Cb m**
mm	Configuration of isotactic triad (i)	
MO	Molecular orbital	
m_p	Mass of the proton	**kg**
$\overleftrightarrow{mr}$	Configuration of heterotactic triad (h = mr + rm)	
MW	Microwaves	
n	Refractive index	
n	Number of monomer units in a polymer molecule	

nIR	*near infrared*	
N_L	Loschmidt's number	**mol^{-1}**
	(or Avogadro's constant N_A)	
NMR	*Nuclear Magnetic Resonance*	
nUV	Near UV	
OD	Optical density or absorbance ($=\log(I_0/I)$)	
P	Polarisation energy	eV
PFT	*Pulse Fourier-Transform* (NMR)	
Q	Quenching factor	
R	Nuclear coordinates and distances	nm
r	Electronic coordinates and distances	nm
r	rocking vibration	
r	racemic (configuration of dyad)	
rf	radio frequencies	
rr	Configuration of syndiotactic triad (s)	
S	Symmetry coordinate	
s	strong (IR absorption)	
s	syndio	
S_0	Singlet ground state	
$S_1, S_2...S_n$	Excited singlet states (relative to S_0)	cm^{-1}, eV
S_n	n-Fold rotation – reflection axis of symmetry	
S/N	*Signal to Noise* (ratio)	
st	Stretching vibration	
st	Syndiotactic	
T	$\left(=\dfrac{I}{I_0}\right)$ Transmission	
T	Absolute temperature	**K**
t	Time	ps, ns, µs, ms, **s**
t	Twisting vibration	
$T_1, T_2...T_n$	Triplet states	cm^{-1}
	(relative to S_0 ground state)	eV
T_1	Spin-lattice relaxation time	**s**
T_2	Spin-spin relaxation time	**s**
T_n	Translation (symmetry element)	
TMS	*Tetra Methyl Silane* (as chemical shift standard in NMR)	
UV/VIS	Mostly used for nUV + VIS	

E. Conclusion and Appendices

v	Velocity	$\mathbf{ms^{-1}}$ cm s^{-1}
v	Quantum number of vibration	
$v_0 \ldots v_r$	Vibrational quanta in the ground state and in excited electronic states	cm^{-1}
VB	Valence band	
VIS	Visible range of the electromagnetic spectrum. light	
vs	Very strong (IR absorption intensity)	
vw	Very weak (IR absorption intensity)	
w	Wagging vibration	
w	Weak (IR absorption and intensity)	
Z	Nuclear charge	nucleus^{-1}

Greak Letters

α	Absorption coefficient, base e; mostly used in solid state physics	cm^{-1}
α	Polarisability	cm^3, $\mathbf{m^3}$
γ	Out-of-plane bending vibration	
Δ	Peak- or band width	eV, cm^{-1}, $\mathbf{Hz}$
Δ	Thermal energy (in chemical equations)	
δ	Escape depth ($1/e$) of electron (ESCA)	nm
δ	Pathlength difference (in Michelson's interferometer)	cm
$\langle(\Delta B)^2\rangle$	Second moment of absorption curve	
ΔE	Energy gap (solid state physics)	eV
δ_i	Chemical shift, relative to internal standard, mostly TMS	ppm
ε	Molar absorption coefficient, base 10; mostly used in chemical spectroscopy (concentration in mol l^{-1})	l mol^{-1} cm^{-1}
ε'	Specific absorption coefficient, base 10 (concentration in gl^{-1})	l g^{-1} cm^{-1}
ε	Dielectric constant	
$[\eta]$	Intrinsic viscosity	dl g^{-1} (1 dl = 0.1 dm^3)
η_F, η_P	Experimental quantum efficiency of fluorescence or phosphorescence	
θ	Moment of inertia	$\mathbf{kg\ m^2}$
$\varkappa^2$	Orientational factor in Förster's equation	
λ	Wave length	nm, μm, cm, $\mathbf{m}$
$\boldsymbol{\mu}$	Dipole moment (electric)	D (Debey, 10^{-18} esu) $\mathbf{Cb\ m}$
μ	Reduced mass (of molecular vibrations)	

μ_B, μ_N	Bohr's magneton, nuclear magneton	$\mathbf{JT^{-1}}$ $(\mathbf{A\,m^2})$
v	Frequency	$\mathbf{Hz\,(s^{-1})}$, kHz, MHz, GHz
v	Stretching vibration	
v'	Wave number	cm^{-1}, $\mathbf{m^{-1}}$
ϱ	Density	$g\,cm^{-3}$, $\mathbf{kg\,m^{-3}}$
ϱ	Spin density (ESR)	
σ	Absorption cross section, base e (concentration in molecules cm^{-3})	cm^2
σ	Specific electric conductivity	$\Omega^{-1}\,cm^{-1}$ $(S\,cm^{-1})$
σ_i	Shielding constant, consisting of a diamagnetic part σ_i^d and a paramagnetic part σ_i^p	
τ	Decay time $(1/e)$ or "life-time" of excited states	ps, ns, μs, ms, s
τ	Torsional vibration	
τ_c	Rotation correlation time (spin labels)	s
τ_e^F	Radiative (fluorescence) lifetime	ms, s
τ_{lm}	Rotation correlation time of segments and side groups (lm = local mode)	s
τ_{tot}	Rotation correlation time of macromolecule (coil)	s
ϕ	Phase difference (phase angle)	$\mathbf{rad}$
ϕ	Work function = ionisation energy of metals or location of E_F below the vacuum level in pure polymers (insulators)	eV
ϕ_F, ϕ_P	True quantum efficiency of fluorescence or phosphorescence	
$\varphi(\mathbf{R})$	Nuclear wave function depending on the space coordinates of the nuclei	
Ψ	Wave function	
$\psi(\mathbf{r})$	Electronic wave function, depending on the space coordinates of the electrons	
ω	Angular velocity	$\mathbf{s^{-1}}$

Appendix 3

List of Elementary Constants Used in This Book[1]

Quantity	Symbol	Value	Units	Uncertainty, parts 10^6
Speed of light in vacuum	c	2.997924580 (12)	$10^8\,ms^{-1}$	0.004
Permeability of vacuum	μ_0	4π exactly	$10^{-7}\,Hm^{-1}$	
Planck constant	h	6.626176 (36)	$10^{-34}\,J\,Hz^{-1}$	5.4
		4.135701 (11)	$10^{-15}\,eV\,Hz^{-1}$	2.6

1 Data according to National Physical Laboratory, Teddington, U.K.

E. Conclusion and Appendices

List of elementary constants used in this book (continued)

Quantity	Symbol	Value	Units	Uncertainty, parts 10^6
$h/2\pi$	h	1.0545887 (57)	10^{-37} J s	5.4
		6.582173 (17)	10^{-16} eV s	2.6
Elementary charge	e	1.6021892 (46)	10^{-19} C	2.9
		4.803242 (14)	10^{-10} esu	2.9
Avogadro constant (or Loschmidt's Number N_L)	N_A	6.022045 (31)	10^{23} mol^{-1}	5.1
Bohr magneton	μ_B	9.274078 (36)	10^{-24} J T^{-1}	3.9
		5.7883785 (95)	10^{-5} eV T^{-1}	1.6
Nuclear magneton	μ_N	5.050824 (20)	10^{-27} J T^{-1}	3.9
		3.1524515 (53)	10^{-8} eV T^{-1}	1.7
Magnetic moment of the electron	μ_e	9.284832 (36)	10^{-24} J T^{-1}	3.9
Magnetic moment in Bohr magneton (g-factor, g_e)	μ_e/μ_B $=g_e/2$	1.0011596567 (35)	–	0.0035
Ratio of electron and proton magnetic moments	μ_e/μ_p	658.2106880 (66)	–	0.010
Mass of the electron at rest	m_e	9.109534 (47)	10^{-31} kg	5.1
1 Electron volt	eV	1.6021892 (46)	10^{-19} J	2.9
in frequency units		2.4179696 (63)	10^{14} Hz	2.6
in wave number units		8.065479 (21)	10^5 m^{-1}	2.6
in temperature units		1.160450 (36)	10^4 K	31
Magnetic moment of free proton	μ_p	1.4106171 (55)	10^{-26} J T^{-1}	3.9
in Bohr magnetons	μ_p/μ_B	1.521032209 (16)	10^{-3}	0.011
in nuclear magnetons	μ_p/μ_N	2.7928456 (11)	–	0.38
Mass of proton at rest	m_p	1.6726485 (86)	10^{-27} kg	5.1
Ratio of proton mass to electron mass	m_p/m_e	1836.15152 (70)	–	0.38
Mass of neutron at rest	m_n	1.6749543 (86)	10^{-27} kg	5.1
Energy x wavelength (or energy $\div$ wavenumber)		1.2398520 (32)	10^{-6} eV m	2.6
Boltzmann constant R/N_A	k	1.380662 (44)	10^{-23} J K^{-1}	32
		0.861735 (28)	10^{-4} eV K^{-1}	32
Molar gas constant	R	8.31441 (26)	J mol^{-1} K^{-1}	31

"Bases of Measurement" is obtainable from the Publications Officers, National Physical Laboratory, Teddington, Middlesex, TW 11 OLW

Author Index and Subject Index

Absorbance   29
Absorption band(s)   23
 , cell (ESR)   120
 , electronic   28
 , integral   28, 77
 , intensity of   27
Absorption coefficient   27, 28
 , molar decadic   29, 31, 77
 , specific   28, 29
Absorption cross section   29
Absorption edges   23, 24
Absorption of radiation
 , general law of in homogeneous matter
 28
Absorption spectroscopy
 , ultraviolet and visible   23
 , near ultraviolet and visible   25, 26
Abbreviations, table of   174
Accordion vibration   56, 70
 , in paraffins   71
Acetophenone type end groups in PS   48
 , C=O valence vibration   48
 , IR absorption of   48
 , phosphorescence of   48
Acoustical branch (of phonons)   60
Acrylic acid   125
Acrylonitrile   125
Additives   33
Adenine   32
Adhesion   21
AFC   120
Amide bands (IR)   89
Auger effect   10
Angular momentum   109, 110, 113
 , velocity   109
Anionic polymerisation   125
Anisotropic broadening   119, 120
Anthracene   39
Anti oxidants   33
Antistatics   33
Anti-Stokes process in INS   61, 103
 , in Raman spectroscopy   63, 65
Ar$^+$ laser   66
Aromatic polymers   34
Atmospheric pollutants   33

ATR   80, 81, 96, 97
Avogadro constant (Loschmidt's number)
 180

Backbone (of polymer)   29
 , poly conjugated   74
Bamford   123
Band
 , pseudo 1D   23
 , gap   24
Band model   16
 , of polymer backbone   18
Bargon   160
Bending (vibration)   83
Benzotrifluoride   11
Benzoyl peroxide (CIDNP)   160
Bergmann   158
Bernoulli statistics   151, 155
Binding energy, of electrons   7
Biological polymers   4
Biopolymers   4, 30, 32, 74, 75
 , ESR   116
Biphenylene   41
Biradicals   123
F. Bloch   15, 111
Block copolymers   41
Bloor   73
Bohr's magneton   110, 114, 180
Boltzmann constant   180
Boltzmann distribution (equation)   81
 , in Raman spectra   63, 65
 , in ESR   119
 , in NMR   138, 142, 143
Born-Oppenheimer approximation   26
Bragg's diffraction (law)   16, 61, 103
Branching (of polymers)   95, 165
 , back-biting mechanism   167
Breathing vibration   64, 83
Bresler   123
Brillouin   15, 16
Broad-line spectra (NMR)   139, 144, 157, 158
de Broglie   16
Broglie wavelength   61, 103, 104
Broglie's formula   61

Bulk analysis 13
, ESCA 14
Bulkin 65

Cameron 135
Cantow 165
Caoutchouc 96
Carbene-type chain end (TSHD) 133
Carbon double bonds
, in IR bands 96
Carbonyl group IR absorption 77, 84, 88, 95
, in degraded films of PE 98
CARS 66, 75
Catalyst 33
Cationic polymerisation 126
Cavalli 156
Cavity (ESR) 120, 121, 124
, scheme 122
CB 15
Centre of symmetry 56
, of 1D chain 57
Chain dynamics 47
Chain segment, movements of 2
Character table 58
Charge transfer absorption bands 33
Chemical shift, ESCA 10, 11
Chemical shift 138, 139, 144
, anisotropy 163
, data 165
Chemical structure 2
Chromophores 27, 77
CIDNP 143, 160
(cis)poly(1,4-butadiene) 72, 96
cis-1,4-poly isoprene 96
Clar 31
Clark 13
C≡N band 95
^{13}C–NMR 161, 165
, of solid polymers 162
, of solutions 165
, résumé 169
Coherent INS 104
Compatibility (of polymers) 47, 51
, excimer fluorescence as a probe 47
Commercial polymers 33
Complex formation 2
Conduction band 16, 19
Configuration(s) 2
, in PP 90
Conformation 2
, in mIR spectra 89
Copolymerisation (ESR) 125
Copolymers 2, 30
, alternating 168
, analysis of 32
, mIR 94
, ^{13}C–NMR 167

, singlet excitons 44
, statistical (random) 41, 82
Core electrons 8–11, 21
Corona discharge 13, 15
Corona treatment (polymer films) 80, 98
Correspondence principle 54
Coupling „electrical" 86
, constant (hyperfine splitting) 118
, "mechanical" 86
Covalent bands 16
CP-MAS 164, 169
"Crankshaft-like" motion 47
Critical radius, R_0 44
Cross linking 98
Cross polarisation 159, 163, 164
Cross-section for neutron scattering 104
, UV/VIS absorption 29
Crystal lattice 18
, infinite 1D 56, 59
Crystal symmetry 15
Crystals, molecular 17
Crystallinity 2
, 1D 2, 18
, 3D in mIR 89, 99
, degree of (Raman) 69
, band 87
CT complexes 76, 126, 135
Cuniberti, Carla 47
CW-NMR, Comparison with PFT-NMR 161
Cytosine 32

DA complexes 33, 100
Davydov splitting 32, 68, 90
Debye (unit) 28, 77
Decay behaviour of luminescence 40
Decay time(s) 40, 41
, of fluorescence labels 43
Deformation (vibration) 83
DGEBA 164
, solid state ^{13}C–NMR 165
Degradation of polymers 34
Delayed emission 36
Delker 76
Density of states 16
Density of unpaired electrons ("spin density")
119, 125
Diamagnetic shielding 139
Diamagnetism 139
Dicarbenes 133
2,4-Dichloropentane, meso 153
, racemic 153
Diffuse reflection 80
Dimeric radical 124, 125
2,6-Dimethyl phenol in PE
, fIR and mIR, H-bond 101
2,4-Dinitrophenylhydrazine for detection of
ketones and aldehydes 99

9,10-Diphenyl anthracene   39
Dipole moment   64, 77
  , induced   64
Dipole resonance   43
Dirac   110
Dispersion interaction   32
DNA   30,41
  , hypochromy   32, 34
  , random coil   32
1D system (chain), dispersion curve for
  longitudinal phonons   60
  , crystallinity in Raman spectrum   69
Dyad(s)   148, 149
Dyes   33
Dynamics of macromolecules   51
Doublet (states)   130
  , spin   110
  , splitting of energy levels in magnetic field
  114
Douglas   165

Eigenvalues (of harmonic oscillator)   54
Electric field distribution in ESR cavity   122
Electromagnetic radiation, spectrum of   3
Electron affinity   17, 27, 33, 42
Electron analyser   7, 9
Electron-nuclear spin coupling   141
Electron spectroscopy for chemical analysis
  (ESCA)   7
Electron-spin polymers   123
Electron spin(s)   26, 27
Electron volt   180
Electronic absorption bands in Resonance
  Raman Spectroscopy   72
Electronic ground state   26, 42
Electron spectroscopy   4, 7, 8
Electronic structure   2
Electronic vibronic spectra   26
Electronically excited state(s)   26
Electrostatic charging   21
Elementary charge   180
Elementary constants, list of   179
Emission spectra
  , time resolved   40
  , "true"   39
Emitting chromosphere   40
End-group cyclization   47
Energy band model   29
Energy bands   15, 16
  , in polymers   17
Energy donor   43
Energy gap   15
  , of HDPE   19
Energy level diagram(s)   27
Energy transfer   42, 43
  , between isolated groups   43

Energy acceptor   43
Escape depth of electrons   13
ESR spectra of mechanically formed
  polymer radicals   126
ESCA   3, 7, 25, 80, 98, 171
  , résumé   21
  , spectra   9, 10, 12
  , spectrometer   8, 9
ESR   4, 47, 109, 111, 135, 136
  , absorption   121
  , general characteristics of spectra   113
  , spectral range   111, 112
  , spectrometer   120, 121, 122
  , spectroscopy of polymers   113, 123
Ethanol   139
  , (NMR)   140, 142
Excimer(s)   42
  , fluorescence   45, 46
  , potential scheme   45
  , formation   45, 47
  , as a probe in polymer studies   46, 47
Excimer-forming sites   45
  , in aromatic polymers   46
Exciplex fluorescence   32
Excitation spectra   39
Excited states   42
  , vibrational, emission from   80
Exciton(s)   2, 25, 45, 50, 135
  , hopping   42
  , in molecular crystals   46
  , wavelike (coherent)   45

Factor group   57
Fenton's reagent   124
Fermi level   18, 19
Fermi resonance (splitting)   68
Fibres   33
Field scan   143
fIR   77, 78
  , résumé   105
  , spectroscopy of polymers   100
  , chemical applications   100
Fischer, H.   123
Flow system for radicalic polymerisation
  (ESR)   124
Fluorescence   4, 37
  , comparison with Resonance Raman
  Scattering   72
  , disturbance of Raman spectra
  , excitation   42
  , from isolated chromophores   42
  , in polymers   41
  , labels   43, 134
  , mirror image   37
  , spectroscopy of polymers   36
vibrational structure

Fluorescence and phosphorescence spectra in comparison to vibrational spectra   53
, low temperature   39
Fluorescent groups   41
, burning out   66
, crowded   41
, isolated   41–43
Fluorescent transitions   37
Fluoro polymers   14, 22
FMIR   80, 81, 96
Forbidden trasitions   36
Force constant   54, 55
Förster   43
, theory, mechanism   44
Fourier transform(ation) interferometers   78
, in NMR   161
Fracture of polymers   123
, PA-6   127
, primary processes   128
Franck-Condon principle   28
Frank, C.W.   47
Free electron ESR   129, 131
Free induced decay   162
FTIR spectrometer   79, 81
Fundamental vibrations   53, 54
, of 1D chain molecules   56
fUV   3, 8, 34
, absorption   25
, experimental set-up   23

Garcia   33
Gaußian absorption curve (ESR)   115
George   24
g-factor ($=g_e$) of free electron   114, 180
, anisotropy   127
g_N-factors of nuclei   111
Goddu   76
Grant   91
Grotthus conductivity   142
Group vibrations (frequencies)   81, 82
, notation   83
Guanidine   32

Half-field resonance (ESR)   131
Half width of ESCA peaks   10, 12
Harmonic oscillator   53, 54
H-(hydrogen-)bonds   74, 86, 99, 100
HDPE   18–20, 23, 24, 166
, Raman bands   67, 71
, Fermisplitting   68
, fIR spectrum   102
Helix   7, 32, 71, 91, 93, 104
Hendra   69
Henniker   82
Heteroaromatic polymers   34
He/Ne Laser   66

High-resolution ^{1}H-NMR (of Polymers)   144
Hindered rotation (of spin label)   134
^{2}H-NMR   169
^{1}H-NMR, résumé   160
Hole burning   119
Hole, positive ($=$ defect electron)   8, 24, 27
Holland-Moritz   94
Holographic gratings   65
Homogeneous line broadening   119
Homopolymers   41
, aromatic   45
, singlet excitons   44
Hopping process (model)   44, 136
, singlet exciton   44
Hot bands   67
Hückel (method)   19
Hummel, D.O.   4, 82, 94
Hummel and Scholl   82, 94
Hydroperoxides   49, 129
, mIR absorption   97
2-Hydroxy benzophenones   33
2-(2-Hydroxyphenyl)benzotriazoles   33
Hyperfine (splitting) coupling   117–120
, acrylic acid   125
Hypochromy   30
, in polymers   32

Identity (symmetry element)   56, 57
Inelastic neutron scattering (INS)   61, 100, 102
, résumé   105
Infrared lines (in comparison to Raman lines)   63
Infrared spectroscopy of polymers   75
, empirical   82
Ingram   123
Initiators   124
Insulator   16, 17
Internal conversion   37
Internal standard   140
Interference pattern   78
Interferogram   79
Inter-system crossing   37
Ionisation energy   7, 9, 10, 11, 17, 19, 27, 33
IR-absorption   78
, dispersion curve   60
, spectroscopy   76
IR emission   80
IR radiation, polarised   90
IR reflection   80
IR region of electromagnetic radiation   76
IR-spectrometer(s)   78
Irreducible representations   57, 59
Isotactic configuration
, excimer formation   46
, definition   147

Isotactic sequence   150
Isotactic vinyl polymer
 , extended chain projection   147
Isotopic substitution   152
Isotropic splitting   117

Jablonski-diagram   27, 36
 , simplified   40
Johnsen, U.   146

Kasha   27
 , rule   37
Kausch   123, 128
Keighley   123
Kinetic law of polymerisation (NMR)   120
Klystron   120
Kobayashi   58
Koopmans theorem   7, 21
Kricheldorf   168
Krimm   87
Kr^+ laser   66

LDPE   18, 23, 100
 , branching, lack-biting   167
 , branching, mIR   95
 , branching, NMR   166
 , Raman bands   67
 , Fermi splitting   68
Lambert-Beer's law   29, 33, 95
 , application to polymers   28
 , in IR spectroscopy   77
Lamellae (of crystalline PE)   70, 71
Laser radiation   62, 63
Laser-Raman spectra   66
 , spectrometer   63, 74
Lattice reciprocal   16
 , symmetry   16
 , vibrations (see also phonons)   93
 , wave number   16
 , constant   56
Light scattering   30
Lindberg   123
Linear PE   158, 166
Linear polymers, mIR spectra   82
Linear PU   168
Line broadening (ESR)   119
Line width, natural   10
Living polymer   125
Local modes   135
Longitudinal acoustic vibration   70
Loschmidt's number (Avogadro constant)
 180
Lubricants   33
Luminescence   4, 25, 36
  , excitation spectra   39

Macromolecules, orientation of   43
Macro radicals (s.a. polyradicals)   128
Magic angle   139
 , spinning (MAS)   158, 161–163
Magnetic dipole interaction   138, 157, 163
Magnetic field distribution in ESR cavity   122
Magnetic moment   113
 , free electron   110, 180
 , proton   110, 180
Magnetic spin-spin interaction   131
Mass of the electron   180
 , of neutron   180
 , of proton   180
Mc Brierty   165
Mc Connell's equations   135
Mechanical degradation   136
Mechano radicals   126, 129
meso placement   148
Methyl group rotation (NMR)   158
Michelson Interferometer   78–80
Microtacticity   147, 148, 160, 165, 168, 171
Microwave absorption of polymers   1
 mIR   4
 , spectra of polymers   77, 81
 , applications   94
Mixing chamber (ESR flow system)   124
Model compounds, monomeric   2, 30, 33, 49
 , dimeric or trimeric   46
Modulation amplitude (ESR)   123
Mössbauer spectroscopy   3
Molar gas constant   180
Molar mass   1
 , delayed fluorescence dependent on   49
 , determination by mIR spectroscopy   95
Molecular dynamics   46
 , orbital   27
Molecules, dissolved in solid polymers
 , absorptioon (UV)   33
Moment of inertia   109
Momentum transfer (INS)   61
Monomer, residual   33
Monomer radical   124, 125
Monomer aspect   3, 29, 55
 , definition of   2
Morawetz   46
Multiplicity (spin)   37

Natta   146, 147
Natural linewidth   116
Neutron properties   103
 , radiation   1
 , scattering experiment (scheme)   103
^{14}N-hyperfine splitting (triplet)   118, 125
 , in spin label   134
NIPCA (N-isopropyl carbazole)   31

nIR   4, 34, 76
Nitroxide(s)   133
   , spin label   134
NMR   4, 109, 137
   , pulse technique   139
   , sensitivity   138
   , spectral range   111
NMR spectrometer, block diagram   143
^{15}N-NMR   168
Norrish I and II splitting   97
Nuclear magnetic resonance spectroscopy
   (of polymers)   137
   , magneton   137
   , quadrupole magnetic resonance   109
   , spin   137
   , of isotopes   138
   , spin coupling   138, 140
   , triplet state   118
nUV   3, 26
   , chromophores   24

OH (starting radical)   124
Okamoto   156
Oligo-ethylene oxide   72
Oligo-TFE   72
Optical branch (of phonons)   60
Optical brighteners   33
Optical density   27, 29, 77
Orbital angular momentum   110
Orbitals   9
   , atomic   7
   , molecular   7
   , valence   9
   , core   9
Oscillator strength   28, 31
Overlap integral (Förster)   44
Oxidation   96
   , degradation   97
   , photochemical, polymer films   80
Ozone   97

P1VN   119
P2VN   47
PA6   46, 77, 88, 127, 134
   , mIR spectrum   88
   , mechanically degraded (ESR)   126
Painter   58
PAN   99
   , mIR spectrum   87
Paraffins Raman bands   70
Paramagnetism   47
Parker   39
Partridge   23
Pauli   110
PB   11, 18

PE   11, 25, 56, 82, 129
   , absorption edge   23
   , as matrix in fIR   100
   , broad-line NMR   157
   , ^{13}C-NMR   167
   , extended chain   58
   , fundamental vibrations   59
   , lamellae thickness   72
   , OOH groups   97
   , Raman spectrum   66–68
Peak-to-peak width (ESR)   115
Pentad   155
Permeability of vacuum   179
Peroxy radical(s)   127, 129
Perrin (Francis), Perrin's model   49
Perrin (Jean)   43
PES   8
PET   12, 13, 25, 98, 101, 102
Peterlin   128
PFT-NMR   169, 191
   , comparison with CW-NMR   161
Phase angle of vibrations (phonons)   59
Phase separation of polymer blends   51
Phenoxy radicals (ESR)   130
Phonons   2, 59, 60, 70, 74, 77, 99, 100
   , in fIR spectra of polymers   101
Phosphorescence excitation spectrum of
   commercial PS   48
Phosphorescence   4, 37
   , spectroscopy of polymers   36, 133
   , in polymers   47
   , sensitized of guest molecules   49
Phosphorescent groups isolated   48
Photochemistry   42
   , of polymers   34, 50
Photoconductivity   21, 50
Photo current   24
   , intrinsic, threshold energy of   24
Photo-dissociation   25
Photo electron spectroscopy   8
Photo emission, of electrons   21
Photo-ionisation   25
Photo polymerisation (TSHD crystals)   133
Photo selection   43
Photo oxidation   13
PIB, mIR spectrum   85, 86, 91
   , ESR   129, 130
Picosecond spectroscopy   40
Plane of symmetry   56
Planck constant   179
Plasma polymers   21
Plasticisers   33
Plastics   33
Platt   31, 32
Plexiglas   82
PMMA   25, 82, 84
   , isotactic   146, 147

, mIR bands and spectrum   84
, NMR data   150
, syndiotactic   85, 145
PMS   151
, st   152
, it   152
POE   25, 134
Point group   57, 58
Polarisability (Smekal-Raman effect)   64
, tensor   64
Polarisation, electronic   18
Poly(acrylonitrile)   87
Poly(acrylonitrile-co-methylmethacrylate)
-mIR Spectrum   94
Polyalkylmethacrylates   47
1,2-Polybutadiene   96
Polybutadiene   168
Poly(caprolactam)   88
Polydiines   32
, single crystals, Resonance Raman spectra
73
Polyester(s)   77, 134
Poly(isobutene)   85
1,2-Polyisoprene   96
3,4-Polyisoprene   96
Polymer aspect, definition of   2, 3, 29, 171
Polymer chains, dynamics of   43, 133
Polymer chromophores   36
Polymer coils   30, 65, 135
, free draining   135
, non draining   135
Polymer crystals (folded chains lamellar
structure)   71, 75
Polymer films, tip-coating   23
Polymeric acids   99
Polymeric DA complexes   34
Polymerisation studies (ESR)   123
, radicalic   125
Polymer spectroscopy   3
, definition of   1
, information obtained by   2
, spectral range   3
Polymer surfaces, Corona and plasma treat-
ment   98
Poly(tetrahydrofuran)   72
Polyolefines   129
Polypeptides   88
Polypropylene, amorphous   85
Polyradicals   111, 124, 125
Poly(vinyl alcohol)   86
POM   18, 82, 104
Potential curves   26
Polymers, degree of orientation in solid   43
, table of   173
PP   11, 18, 33
, mIR spectrum of atactic   85, 86, 91
, isotactic   89, 91

, syndiotactic   90–93
, OOH groups   97
, Raman spectrum   66, 69
Primary processes (ESR)   136
Primary radical (ESR)   127
Prompt fluorescence, relation to delayed
fluorescence   49
Proteins   30
, IR/Raman   89
Proton exchange (hopping) processes   142
Proton spin moment   137
PS   11, 25, 30–32
, escimer fluorescence   45
, monomeric models   29
, spin label study   134, 135
, technical grades   30
, phosphorescence spectrum of
, triplet energy   47
PS/Poly(vinylmethyl ether) blends   47
PS/VCA (copolymer)   48
PTFE   13, 14
, Raman spectrum   66, 69
, fIR   102
, INS   104
PTS   32
Purcell   111
PVCl$_2$ 59, 93
PVAC   25
, syndiotactic   86, 87
PVC   11, 30
, branching   167
, double bonds   96
, ESCA valence spectrum   21
, NMR spectrum   154
, radicalic, NMR   153, 155
, tacticity   152, 153
, tacticity analysis   155
PVCA   25, 30, 31, 136
, excimer fluorescence   45
, NMR   156
, triplet energy   47
PVF   11
PVF$_2$, fIR   101, 102
PVOH   11, 13, 99
, mIR spectrum   86, 87

Q band   114, 120
, resonance conditions   122
Quantum efficiency   40, 50
, experimentally observed   40
, "true"   40, 41
, fluorescence   39
Quenching   39, 50
, experiments   40
, factor   45, 46

Quinine sulphate 39
Quintet states (ESR) 133

Rabek 117, 123
Racemic placement 148
Radiation, electromagnetic 1, 112
, induced ionic polymerisation 126
, less transition 36
, particle 1
Radiative transition 36
Radical, anions 119
, cations 136
Radical-ion complexes 33, 76
Radicalic polymerisation (PMMA) 151
Radical, formed by radiation (ESA) 129
, CIDNP 160
Raman spectra 57, 62
, intensity 65
Raman spectrometer 63, 65
Raman spectroscopy 1, 53, 62
, resonance 53
, of crystalline PE 68
Raman spectroscopy (of polymers) 62
, résumé 74
Rånby 117, 123
Randall 168
Rate constants 40, 41
, of emission 37
Rayleigh equation 30
, scattering 62
Reabsorption 39
, of fluorescence 40, 50
Reduced mass (of oscillator) 54
Reference spectra (IR) 81
, substances (compounds) 40, 78
Regularity bands (IR) 81, 82, 89
, ester stretching in PMMA 84
Resins 2, 82, 141
Resonance λ (ESR, NMR) 112
Resonance absorption 110
, condition 111
, frequency 137
, of proton 138
, NMR 138, 144
, of unpaired electron 114
Resonance Raman scattering 32, 72
, of PTS single crystals 73
RNA 30
Rocking (vibration) 83
Rotation-reflection axis of symmetry 56

Sample, position in cavity (ESR) 122
Saturation of signals (ESR) 119
, at high MW power 121
, in NMR 142

Savoisky 111
Scattering (of radiation) 40
Schaufele 67
Schrödinger equation 54
Scintillation counting 51
Scissor (vibration) 83
Secondary radical (in polymer fracture) 126
Second moment (of absorption curve) 157, 158
Selection rules 27
, in vibrational spectroscopy 53, 55
, of IR-absorption 60
, of Smekal-Raman 63
Semiconductor 16, 17
Sensitivity (ESR) 121
Sequence length (tacticity) 150
Shake-off 9, 10
, up 9, 10
Shielding (NMR) 156, 166
Side groups, movements of 2
Siesler 94
Sillescu 169
Singlet oxygen 49, 97
Singlet state(s) 26
, spin 110
Sixl 133
Smekal-Raman disturbance by fluorescence 66
, effect 39, 62–65, 75
, explanation 63
, increase of intensity with frequency 63, 66
, lines, polarisation 63
S/MMA (copolymer) 42
Snyder 94
Sodium salicilate 23
Sohma 128
Solid-state polymerisation 126
Space group 57
Spectrofluorimeters 37, 38
Spectroscopy, definition of 1
, information obtained by 2
Speed of light 179
Spherulites 65
Spin decoupling 153
, density (density of unpaired electrons) 135
Spin of elementary particles 109, 113, 122
, labels 113, 118, 123, 133
, orbit coupling 116
, probes 133
, quantum number 130
Spin-lattice relaxation time (T_1) 116, 159, 162
, of ^{13}C atoms 163
, of protons 142
, traces of paramagnetic impulses 160

Spinning frequency (MAS) 164
Spin relaxation times 159
Spin resonance spectroscopy 4, 109
, in polymers 111
, spectral range 112
Spin selection rule 37
Spin-spin coupling constant 142, 144
Spin-spin relaxation time (T_2) 116, 159
Stabilisers (UV) 130
Starting radical 124
Steric hindrance (in NMR) 156
Stern-Volmer equations 45
Stokes (lines in Raman spectrum) 63, 65, 66
Stretching experiments (ESR) 128
"Stripping off" (side groups) 129
Strobl 67, 71
Submicro cracks 128
Supraconducting coils 143
Surface analysis 13
, ESCA 14
, layer 8
, modification 80
Symmetry axes (of 1D chain) 57
, plane (of 1D chain) 57
Synchrotron radiation 25, 171
Syndiotactic vinyl polymer, extended chain
projection 147

Tacticity 2, 50, 99, 146, 160
, analysis (NMR) 145
, in mIR spectra 89
Tadokoro 58, 87
Teflon 12
Termination 124
Tetrafluoroethylene, copolymer with ethylene
14
Tetramethyl dioxetane 47
Time-of-flight spectra (neutron scattering)
104
TMS 139
Törmälä 123, 135
Torsion (vibration) 83, 100
Transition dipole moment 27, 28, 43, 47
Transition metal ions 113
Translation (symmetry element) 57
Triad(es) 148
, probabilities 149
Triplet excimers 49
, as exciton traps 49
, in aromatic homopolymers 49
Triplet excitons, in polymers 49
, hopping rate 49
, ESR 135
Triplet state(s) 47, 123
, spin 110
, ESR absorption 130
, decoupling 132

Tryptophane 30
TSHD 74
, ESR of growing chain ends 132, 133
Twisting (vibration) 83
Tyndall scattering (effect) 62, 65
, comparison with Smekal-Raman effect
63, 72
Tyrosine 30

Uncertainty principle 115
, (Heisenberg) 10
UV absorbing polymers 30
, absorption 30
, excitation of Raman spectra 66
, radiation 97, 129
, stabilisers 33
, screening type 33
UVPS 8
UV/VIS 34
, absorption spectra 30
, résumé 34
, absorption processes 36

Valence band 16, 18, 19
, 1D 21
, electrons 9, 10, 11
, vibrations 83
VB 15, 16, 20, 23
, CB transition 24
VCA/S (copolymer) 42
Van Vleck 157
Vibrational spectra interpretation 55
, spectroscopy 4, 53
, structure of electronic bands 53
Vibrations of polymers 53
, totally symmetric 57, 64, 76
, in Smekal-Raman effect 63
Vinylidene end groups 24
VIS 4
VN/S (copolymer) 42

van der Waals (forces) 17, 18
Wagging (vibration) 83
Ward 91, 160
Water-soluble polymers 75
Wave function symmetry or antisymmetric
behaviour 58
Wave guide 120
Wegner 74
Winnik 47
Work function 16, 17

X band   114, 119, 120
  , ESR spectrometer (scheme)   121
  , resonance conditions   122
X-ray fluorescence   10
XPS   8
X-ray small-angle scattering   72, 128
X-rays (radical formation)   129

Young's modulus (elastic constant)   70

Zbinden   96
Zero field levels in symmetrical molecules
  132
Zero field splitting   131, 133, 141
  , of triplet states   132
  , parameters D and E   131
Zhurkov   128
Zig-zag chain (extended)   56
  , accordion vibration of   70
  , isotactic PMMA   147
  , syndiotactic PMMA   148

Spectroscopy: NMR, Fluorescence, FR-IR

1983. 40 figures, 5 tables. Approx. 170 pages
(Advances in Polymer Science
Fortschritte der Hochpolymeren-Forschung,
Volume 54)
ISBN 3-540-12591-4

Contents:

E.D.v.Meerwall: **Self-Diffusion in Polymer Systems, Measured with Field-Gradient Spin-Echo NMR Methods.** This article reviews practice and theory of self-diffusion in polymer systems as measured with the field-gradient NMR methods. Polymer diffusion in dilute and concentrated solutions and melts, and penetrant and diluent diffusion in polymers, are covered together with theoretical interpretation of experiments. The aim of this review is to familiarize workers in the polymer field with these techniques for measuring self-diffusion, and with their applications and benefits. (74 ref.)

C.W.Frank, S.N.Semerak: **Photophysics of Excimer Formation in Aryl Vinyl Polymers.** Fluorescence techniques are powerful methods for obtaining detailed information on the molecular structure of biopolymers and synthetic polymers. Excimer fluorescence is a significant feature of the luminescence behavior of virtually all aryl vinyl polymers. The objective of this review is to concentrate on two aspects of the photophysics of these synthetic polymers and copolymer variants: excimer formation and singlet exciton migration. The article clarifies both the power and limitations of the excimer as a molecular probe of polymer structure and dynamics. (220 ref.)

J.L.Koenig: **Fourier Transform Infrared Spectroscopy of Polymers.** This review covers the theory and application of Fourier transform infrared spectroscopy to the characterization of polymers. The basic theory, the sampling techniques and the spectral operations are described. The applications discussed include the study of polymer reactions, polymer structure and dynamic effects. (344 ref.)

Light Scattering from Polymers

1983. 74 figures. V, 167 pages
(Advances in Polymer Science
Fortschritte der Hochpolymeren-Forschung,
Volume 48)
ISBN 3-540-12030-0

Contents/Information:

W.Burchard: **Static and Dynamic Light Scattering from Branched Polymers and Biopolymers.** The properties of synthetic polymers and biological macromolecules are largely determined by their shape and the internal mobility, which can directly be investigated by Static and Dynamic Light Scattering. The author analyses branched molecules in dilute solution, starting with regularly branched structures and ascending to the randomly branched ones. For the first time he combines the static and the more recent dynamic light scattering methods, thus obtaining new structure-sensitive parameters. As he also combines the different concepts of the traditional polymer scientist and the theoretical physicist, both groups can utilize his results. He investigates the applicability of the Flory-Stockmayer Theory and the further developed Cascade Branching Theory, compares in detail theory and experiment and illustrates his statements with graphs and schemes.

G.D.Patterson: **Photon Correlation Spectroscopy of Bulk Polymers.** The use of Photon Correlation Spectroscopy to study the dynamics of concentration fluctuations in polymer solutions and gels is now well established. The article reviews the development of this field and focuses on the dynamics of fluctuations near the glass transition in bulk polymers. The theory of dynamic light scattering from pure liquids is presented and applied to polymers. The experimental considerations involved in this application of Photon Correlation Spectroscopy are discussed.

Springer-Verlag Berlin Heidelberg New York Tokyo

H. Janeschitz-Kriegl

Polymer Melt Rheology and Flow Birefringence

1983. 144 figures. XV, 524 pages
(Polymers Properties and Applications, Volume 6)
ISBN 3-540-11928-0

Contents: Survey of Experimental Results. – Quasi-Molecular Phenomenological Theories. – Prospects for Predictions on a Molecular Basis. – Industrial Applications. – Appendix A: Linear Visco-Elasticity. – Appendix B: The Time-Temperature Superposition Principle. – Appendix C: The Measurement of Birefringence Effects. – Subject Index.

This work presents a comprehensive review of the empirical behavior of polymer melts, demonstrating for the first time the most recent molecular theories for describing this behavior. The technique of the measurement of flow birefringence is shown to be a useful tool for the investigation of rheological properties of polymer melts. The monograph is intended as an introduction into this new area of polymer science for industrial and university polymer scientists in general and rheologists and process engineers in particular. Graduate students are also addressed. The review is a fortunate combination of experimental and theoretical aspects, clearly arranged and didactically well presented.

J. Štěpek, H. Daoust

Additives for Plastics

1983. 54 figures. XI, 243 pages
(Polymers Properties and Applications, Volume 5)
ISBN 3-540-90753-X

Contents: Introduction. – Additives Which Modify Physical Properties: Plasticizers. Lubricants and Mold-Release Agents. Macromolecular Modifiers. Reinforcing Fillers, Reinforcing Agents, and Coupling Agents. Colorants and Brightening Agents. Chemical and Physical Blowing Agents. Antistatic Agents. – Antiaging Additives (Antidegradents): Difficulty Stabilizable and Nonstabilizable Factors Provoking Plastic Degradation. Heat Stabiliziers. Antioxidants and Metal Ion Deactivating Agents. Ultraviolet Protecting Agents. Flame Retardants. Biocides Against Biological Degradation of Plastics. Brief Survey of Methods Used to Incorporate Additives into Polymer Matrices. – Appendix: Abbreviations and Symbols. – Subject Index.

Here is a practical, contemporary discussion of the most important substances used to modify and improve the physical properties and anti-ageing characteristics of polymer-based materials. Each chapter explores a particular type of additive based on its definition, structure, and classification according to main effects on polymeric materials. These chapters take the reader from the fundamental theory through practical commercial applications for the wide range of additive types. The numerous references (650) included provide a valuable source for expanded study.
More than an essential reference, this volume belongs on the shelves of all students, technologists, and researchers of plastics technology.

Springer-Verlag
Berlin
Heidelberg
New York
Tokyo